BUILDING THE
OVERCRANK ENGINE GEORGINA

BUILDING THE

OVERCRANK ENGINE GEORGINA

Tubal Cain

Model & Allied Publications, Argus Books Limited

ISBN 0 85242 747 0

Model & Allied Publications
Argus Books Ltd.,
14 St. James Road, Watford,
Hertfordshire, England.

The text of this book is based on articles from
the *Model Engineer*.

Printed and bound in Great Britain by
Pindar Print Ltd., Scarborough.

CONTENTS

GEORGINA
A XIXth Century Overcrank Engine

This is a model of a typical vertical engine of the 1850's—in fact, what we now call a 'vertical' was known in those days as an 'inverted vertical'; it seemed natural for the cylinder to sit on the floor and it was not till James Naysmith invented the steam hammer that designers put the cylinder up in the air and the crank on the ground. The arrangement was very common, though not always as decorative; often there was only one 'arch', the other end of the crankshaft being supported in a bearing in the wall. Some designs used a couple of vertical columns supporting a horizontal beam to carry the bearings, whilst at least one had a 'Gothic' arch set in the wall to look like a church doorway with the engine in the middle!

There is no cross-head and guide as is now the rule, the piston-rod itself being guided by a bridle across the columns. Some earlier designs of this type used a sort of parallel motion like a beam-engine, but this type of guide, with a forked connecting rod, was quite adequate as steam pressures were only about 30 lbf/sq. in., the connecting rod was relatively long and side thrust was small. (30 lbf/sq. in. ranked as 'high' pressure compared with the 2 to 5 lb. used by Boulton & Watt).

I have designed the engine to use readily available castings—those from the 'Williamson' Engine, marketed by Stuart Turner Ltd. of Henley on Thames. The casting numbers are marked on the drawings and also mentioned in the text. Stuart's can also supply the various screws and studs needed and it is probable they will make up the casting set and screws, together with the eccentric rod, as a little package. I recommend that when ordering this you also order say 6 in. × 6 in. of 'Oakenstrong' jointing material; it is far better than paper, and a piece that size will last you quite a time. Other materials can be obtained where you wish, but the advertisers in the 'Model Engineer' (including Stuart's) can supply all if need be.

You can add to the engine to elaborate at will; not shown on the general arrangement is a boiler feed-pump—this I describe at the end of the book, and you can fit it or not as you wish. You can fit little lubricators to the shaft bearings, or use oil-holes. And to take it to extreme, you could perhaps one day design and make a governor for it! But as shown on the General Arrangement (Fig. 1) she is a very pretty engine indeed, and runs as well as she looks! So, to the construction.

Arch-heads (Fig. 2)

There are three ways of making these in the absence of castings: (a) Use a brass ring, (b) Bend a piece of steel to the required radius, (c) ditto with a piece of brass. Messrs J. Smith & Sons, 42 St John's Square, London EC1P 1ER, can supply such a ring (though it's a bit costly as to be expected) so let's start with this method. Mark out centrelines at right angles across the faces after having first trued them up by skimming both sides in the lathe. (Just chuck in the 3-jaw and take a cut across; the $\frac{1}{4}$ in.

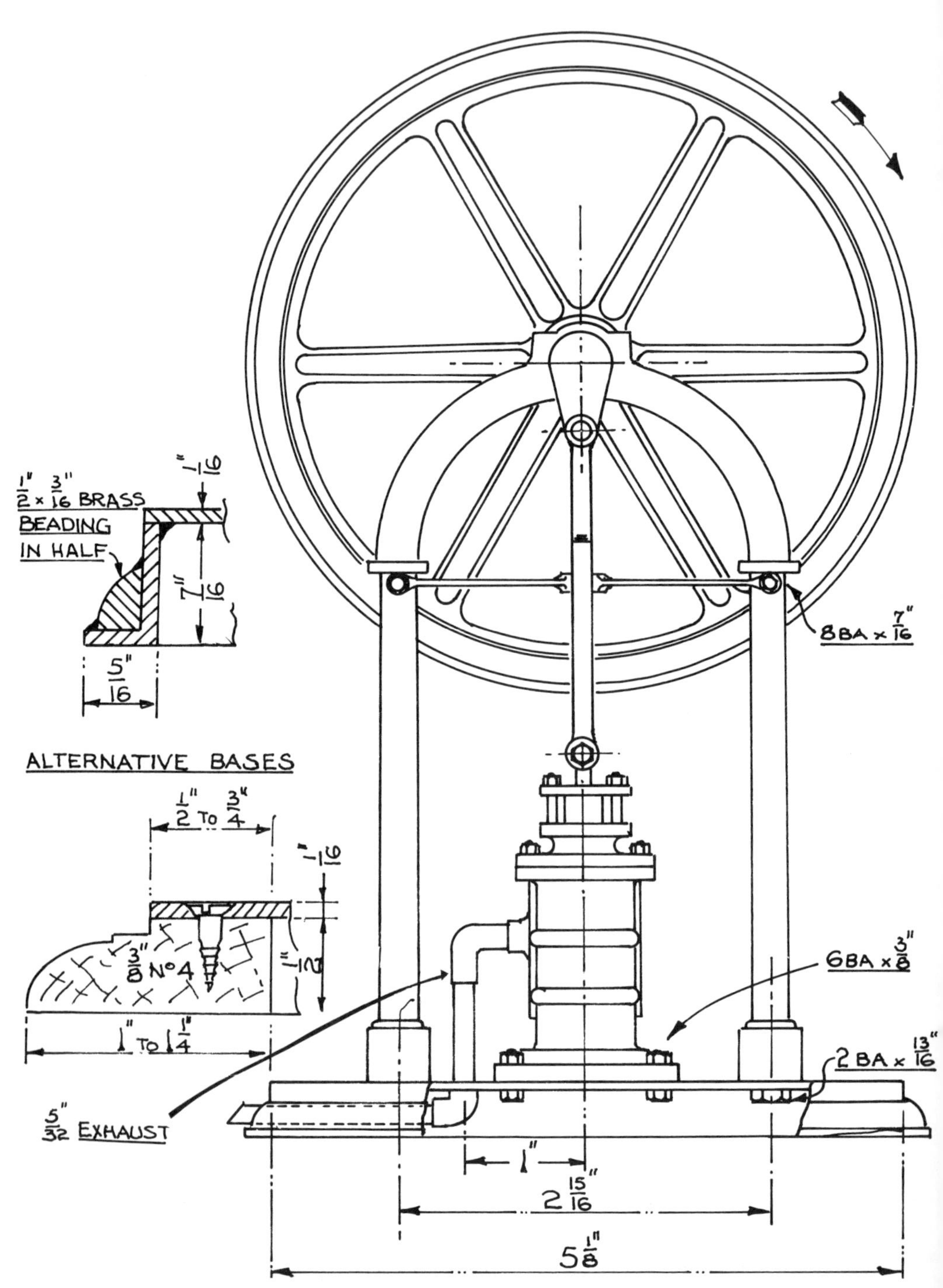

$\frac{1}{2}$" × $\frac{3}{16}$" BRASS
BEADING
IN HALF
$\frac{1}{16}$"
$\frac{7}{16}$"
$\frac{5}{16}$"
ALTERNATIVE BASES
$\frac{1}{2}$" TO $\frac{3}{4}$"
$\frac{1}{16}$"
$\frac{1}{2}$"
$\frac{3}{8}$" No 4
1" TO 1$\frac{1}{4}$"
$\frac{5}{32}$" EXHAUST
8 BA × $\frac{7}{16}$"
6 BA × $\frac{3}{8}$"
2 BA × $\frac{13}{16}$"
1"
2$\frac{15}{16}$"
5$\frac{1}{8}$"

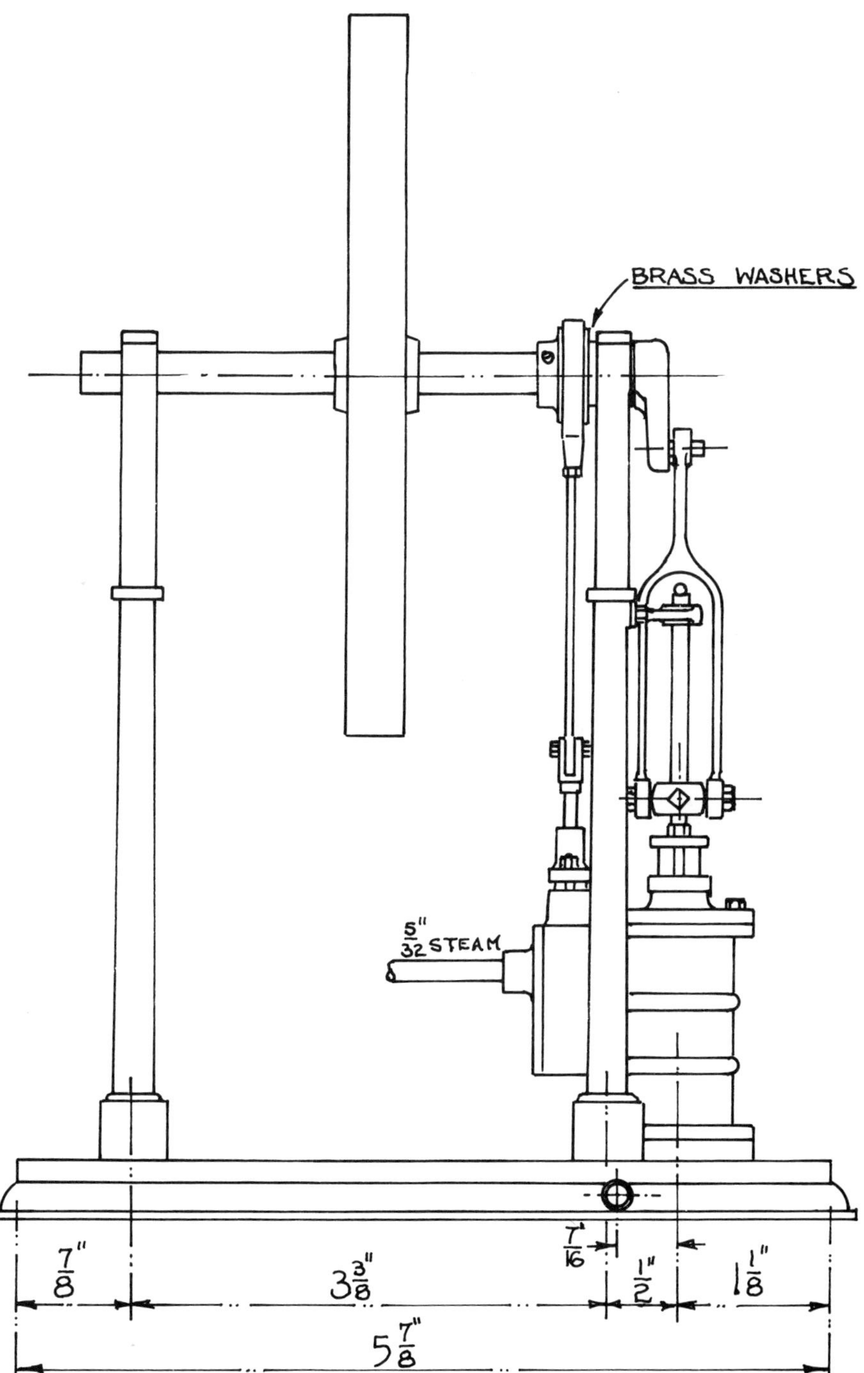

Fig 1-General Arrangement of the Engine.

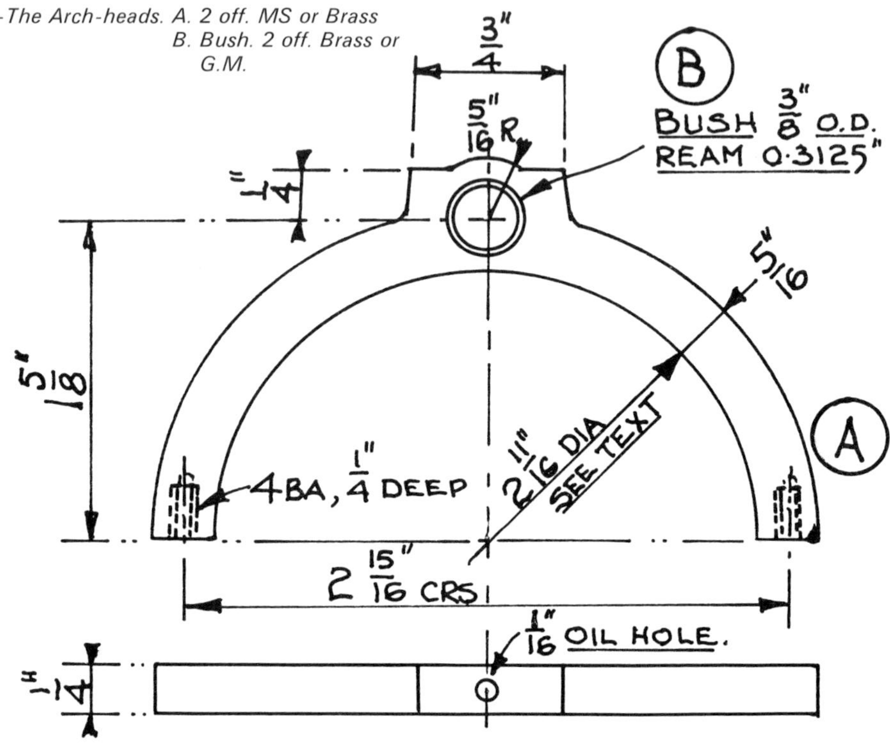

thick isn't critical to $\frac{1}{64}$ in. either way). Then drop a couple of pieces of angle into the vice, grip the ring across one of the diameters you have marked and gently squeeze in till after releasing the pressure the ring is now an oval, $3\frac{1}{4}$ in. across the short diameter. This squeezing gives you the necessary extra on the other diameter to enable you to saw it in two and still maintain the required dimensions.

So, saw the ring across the other diameter, and look twice to make sure you are sawing across the narrow one! From now on the procedure is the same as for the other methods, so let us deal with them now, after which all can march forward together.

Whether you use steel or brass is up to you. I prefer steel, both because it's cheaper and because I find it easier to bend. On the other hand, brazing the little bearing cap on the top is easier with brass. The difference, brass is annealed and bent cold, steel is bent hot. Now, the method I am going to describe produces a helix rather than a ring, but don't worry, it comes out right in the end. Find a bit of stout pipe or cob end of bar which is about $2\frac{5}{8}$ in. O.D. Adjust the vice-jaws to accept this 'round thing' plus just over $\frac{5}{16}$ in., and arrange some packing of some sort to support the round whilst you are busy elsewhere. Get things set so that when you bring the hot piece of steel to the vice you can quickly clamp it up with one hand whilst holding the rectangular steel in tongs with the other. You will need a piece of $\frac{1}{4}$ in. $\times \frac{5}{16}$ in. steel at least a foot long, preferably 18 in. Get the first 6–8 in. red-hot—really red—and if the other end can be held in the hand, well and good; if not, use a mole-wrench. Rush the bar to the vice, get a good hold of the very end between the vice-jaws and the round block (grip towards the very end of the vice-jaws, too) and then as quickly as may be bend the rod round the mandrel; assist progress with light blows from a medium hammer if need be, but don't worry if it isn't absolutely round or is a bit crooked. Return to the heat and whilst the next few inches are redding up, readjust the mandrel in the vice so that you will be able to bend still further. If you have got a 180° bend in at first heat, well and good; next time grip the dead end at the front jaw with the 'handling end' standing up at the back. But if you managed only $\frac{1}{4}$ turn or so, repeat as before. Repeat as necessary until you have just over a complete turn like a close-coiled spring with short length of straight at one end and a longer straight at the other.

Now start the correction process, first concentrating on bringing the coil to circular shape of the right dimensions. Get all hot again, and close up the coil till side is touching side. Set the mandrel projecting from the vice. I fixed mine upright (it was a bit of $2\frac{1}{4}$in. BSP steam pipe) and this worked okay. Reheat and tap here and there with a hammer to correct any ovalities. This will tend to open up the coil, so from time to time you may have to reheat and close it up again. However, not to worry if you are say $\frac{1}{16}$in. out on the dimension; all you have to do is to adjust the centres of the columns on the base and bridle to suit, and make the columns that much shorter or longer to compensate. You will finish off with a close coil of the right dimensions, and probably a nice blacksmith's finish at that. Now saw off the surplus so that you can close the coil axially to make a ring-with-a-slit-in-it. Do this hot, too. Finally, with a fine saw, cut into two half rings, exactly opposite the first saw cut. NOTE: You could, of course, forge two half rings if you like, and if you are reasonably adept with blacksmith's work this is a better way.

For the brass method, the procedure is the same but you must repeatedly anneal the brass and do the job cold. It is necessary to be more careful, as the brass will show both hammer and vice marks much more readily than will steel, and this makes for more trouble later when painting. Don't try to do it hot, as the brass will almost certainly collapse on you unless it is the correct hot-working variety.

I have said three methods, but there is, of course, another very expensive way; that is to band-saw the arches exactly as to drawing from $\frac{1}{4}$in. thick material. In this case the bearing housings can be in one piece if you wish. For the others, the next step is to braze on the additional piece.

File up a block to the shape shown in Fig. 2 and prepare a mating flat on the arch—this flat about $\frac{3}{32}$in. longer than the $\frac{7}{8}$in. dimension; this will give you a nice rounded fillet of brazing material. In a job like this I always relieve the mating faces in the middle by about 5 thou, so that the block sits down at the ends but leaves a small gap in the middle, thus ensuring good penetration of the alloy, but don't make the gap more than that. Flux well, and clamp the two parts together. (The little forged steel clamps available from advertisers are ideal for this purpose.) Easyflo No. 2 is quite in order for this joint. Heat up quickly, apply sufficient alloy to fill the joint and produce fillets at the ends, and pickle. File off the surplus alloy at the sides and trim up the fillets so that they match. Fig. 3 shows two steel arches after brazing; in this case I wasn't satisfied with the fillet—the corner was quite sharp—so after pickling I filled this in with ordinary solder.

We must now machine up the faces which bear on the columns and it is fairly important that these be square otherwise they will sit crooked on the columns. Fig. 4 shows the method. Match up the pair of arches and clamp to the lathe cross-slide on packing. Run the cutter across to trim the ends, paying reasonable attention to the $1\frac{5}{8}$in. dimension. This done, mark out for drilling and

Fig 3-Steel arches after brazing on the bearing section.

Fig 4-Machining the arch footings.

reaming the $\frac{3}{8}$ in. hole for the bushes. You should still be able to see the scribed centreline you put on at the beginning—especially if you followed the proper drill and dotted it with light centre-punch marks. Set up vertically and use your scribing block to mark out the position of the hole. If your drilling machine drills truly square, use this; if not, then you must mount the arches on the faceplate and drill from the lathe tailstock. Do both at once, taking great care that the faces you have just machined are in line. Drill Letter U or 9.3 mm (9.4 leaves less on for reaming) and then follow with $\frac{3}{8}$ in. reamer. Remember, if the arch is of brass, to use a straight-flute drill or hone off the rake angle at the cutting edge of a twist drill, otherwise it may walk its way through and leave a ragged hole.

Mark out the flat feet of the arches with two diagonal lines, centre-pop at the intersection and drill your favourite tapping size for 4-BA, about $\frac{5}{16}$ in. deep. Take care that these holes are square; I set up in the vice on the drilling machine and used my scribing block to ensure that both feet were level. Tap the holes, and check with a piece of studding that the holes ARE square. If they aren't, what then? You *can* try plugging and redrilling, but its 50p to a used stamp that the plug will come out, so the only alternative is to drill for 2BA, fit a plug this size with Loctite and try again. Alternatively, you can just work the tap against the slanting side and try to straighten the hole, which gives a loose fit. That's not too important if you use Loctite on assembly.

The bushes are a plain turning job, but don't ream the hole till you have fitted them if you rely on press-fit. I made mine a drop-in fit and used Loctite, thus enabling me to ream in the lathe and get a better finish. Finally, drill for a $\frac{1}{16}$ in. oil-hole. The purists will say 'why not a loose bearing cap'? You can if you like, but if you do you must braze in a piece on the inside of the arch, to give enough metal to carry the strain. Instead of a loose cap you can drill and tap for two say 8BA studs and nuts to make it look

as if the cap *were* detachable. The point here is that this isn't a model which is exact to a prototype, but a *representation* of a general type of engine, so that refinements like this can be cosmetic rather than functional. A bush will last for years of work before replacing (the shaft of 'The Lady' ran in the brass arch with no bush at all originally).

Columns (Fig. 5)

These may be of steel or brass, and are turned from $\frac{3}{8}$ in. dia. stock. Part off four pieces exactly 4 in. long, and then centre, drill, and tap the holes in both ends of each—2BA one end and 4BA the other. Mark the 2BA end of each with a felt pen or a spot of paint. Recentre each end so that there is a bearing for the lathe centres in each tapped hole, but don't go so deep that the centre at the small end would obliterate the $\frac{3}{16}$ in. dia. spigot. Set up between centres and machine from the large end to $\frac{11}{32}$ in. dia. as far as you can; don't bother about the bit under the carrier—it will come off later. Aim at a good finish and do all four, one after the other. Replace in the lathe and with a knife tool form the $\frac{1}{4}$ in. dia. spigot on all four. Reverse the position of the carrier (taking care not to burr the work) and form the $\frac{3}{16}$ in. spigot at the other end; again, do all four one after the other. This method means that once you have the first to size you simply use your feed-screw index to get the same dimension on each—saves time. Having got so far, carefully compare the four columns and make sure that the $3\frac{27}{32}$ in. dimension is the same on each. It doesn't matter to within 10 thou. but they should be identical.

Now for the taper. This is a bit too long to turn by setting over the top-slide, so set over the tailstock towards you by about $\frac{1}{32}$ in. for trial. Mount the work between centres, carrier on the large end, and first chew off the surplus $\frac{3}{8}$ in. dia. part you couldn't remove previously. Then take a trial cut and check whether the taper will run out at the right place; it doesn't matter a great deal if the $2\frac{3}{4}$ in.

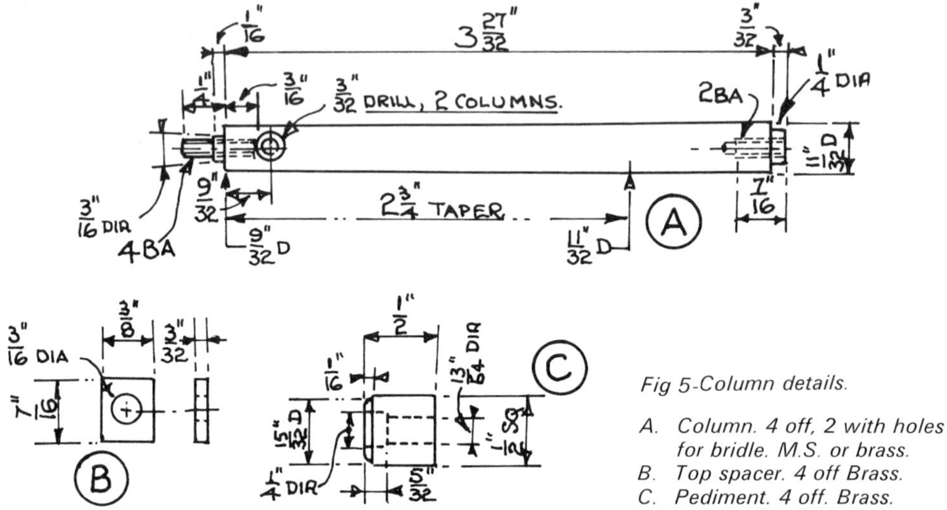

Fig 5-Column details.

A. Column. 4 off, 2 with holes
   for bridle. M.S. or brass.
B. Top spacer. 4 off Brass.
C. Pediment. 4 off. Brass.

is +/−⅛ in. but try to get it right. Once you are satisfied, use power feed and turn the taper on all four columns. Finally, with an emery stick blend the transition from parallel to taper. I recommend you sweep medium coarse emery up and down the length to provide a key for the paint.

The square footings, 3C, are made by chucking a piece of ½ in. square stock about 2¾ in. long in the four-jaw and setting it true. Drill ¼ in. for $\frac{5}{32}$ in. deep, recentre and drill $\frac{9}{16}$ in. deep with a 3 mm drill. Form the little $\frac{1}{16}$ in. radius—I use a form tool for this, made very simply by first grinding and then honing with a tiny carrot-shaped slipstone on the very corner of a standard $\frac{5}{11}$ in. square HSS tool-bit. Finally for this one, part off as accurately as you can to ½ in. long. Repeat this process four times, but don't forget to face the newly exposed end each time. Now, you will at some stage have to feed the work forward out of the chuck. Here is a tip; paint two of the chuck-jaws with yellow paint or something, and always adjust on these two. If you do this you can unchuck and rechuck quite awkward shapes and retain the setting pretty closely.

The $\frac{3}{32}$ in. rectangular 'washers', 3B, for the column top are made by first filing or milling a piece of stock to $\frac{3}{8}$ in. × $\frac{7}{16}$ in., setting this true in the 4-jaw, drill $\frac{3}{16}$ in., and part off four bits. If you haven't parted off square or rectangular stock before, don't worry. It's not difficult; make sure the tool is sharp and at centre-height. The front face should be square (or parallel, depends how you look at it!) to the work, not tapered as some recommend, and the tool should have about 10° of rake for steel, zero rake for brass. Feed in gently until the tool is cutting full circle, then go as fast as you please.

Having done all this, reset the tailstock to turn parallel. I should have told you about this before you did all that drilling but if you forgot (like I did) it won't hurt, as none of the holes is deep.

Now you must check that the columns with their footings and top washers are all the same length. If you have a 6 in. vernier caliper this is easy; just assemble each and check the overall dimension, which should be 4$\frac{7}{16}$ in. As before, a few thou. up or down won't matter so long as they are all the same. I doubt if they will be, though, so what you do is to chop and change columns which are long with footings which are small and so on till they are nearly the same, and finally, carefully file the footing of the long one to bring it down to the others. If you haven't a vernier, then you can work to a thou. or so with your scribing block. Clamp the columns to some support on the lathe-bed, surface plate, or the table of your drill and adjust the point of the scriber so that it just brushes one of them. Then offer it to the others. You will be able to detect as little as 2 thou. this way. Just one point, though; I suggest you put a 4BA screw and a washer to hold the top piece on—it's not heavy enough to stay put by itself. Once you have all four matched up, mark them so that they can be mated later.

There are now two $\frac{3}{32}$ in. holes to cross-drill on two columns only, to carry the connecting-rod guide bridle. Note here that if you have adjusted the column length to allow for a difference in the arch-head radius, these holes must be the same distance from the *bottom* of the column. That is, $3\frac{27}{32}$ in. minus $\frac{9}{32}$ in. = $3\frac{9}{16}$ in. You will almost certainly find that the $\frac{3}{32}$ in. holes don't line up if you drill them blind, so assemble the selected columns to their arches and mark mate to mate. At the dimension given above, file a little flat and mark out the position of the hole. Lay the assembly flat on the table of the drilling machine and carefully—very carefully—drill through. You can't use a jig for this job—well, I suppose you could, but it's a lot of labour just for two holes and you have to take care with the jig anyway! Make sure the hole goes (a) straight and (b) through the middle of the column. Having done this I suggest you offer up both sets of columns to a temporary base and adjust till a piece of $\frac{5}{16}$ in. ground rod passes through both bush holes. I did this with a touch of Loctite nut-lock on the 4BA threads which allowed some manoeuvring but held things in place once all was in order (Fig. 6).

Crankshaft (Fig. 7)

This is a relatively simple shaft, the web requiring more 'work by eye' to make it look well than anything else. If your 3-jaw is reasonably true it is only necessary to chuck a piece of ground mild steel $\frac{5}{16}$ in. OD and after facing both ends turn down one end to $\frac{1}{4}$ in. dia. $\times \frac{1}{4}$ in. long plus say $\frac{1}{32}$ in. for rivetting if you feel you must, though high-strength Loctite will do all that is needed. For the web, select a piece of $\frac{1}{2}$ in. $\times \frac{1}{4}$ in. bright MS a shade over 1 in. long, mark out centre-lines and the position of the two holes $\frac{9}{16}$ in. apart. If your drilling machine (*and* the drilling vice) is true, then you can drill and ream the two holes without further ado. The centre-distance will be okay if within 10 thou. but the holes should be as parallel as may be, or the connecting rod may wag about on the pin when running.

If you don't trust the drill, then you must set up in the 4-jaw and drill that way, but, quite frankly, I would rather trust any but a really ropy drilling machine than try to get two holes parallel at separate settings in a 4-jaw. The third way is to hold the web in a vice on the cross-slide, drill and ream the $\frac{1}{4}$ in. hole from the 3-jaw, traverse over on the cross-slide $\frac{9}{11}$ in. and drill the small hole. This way you *will* be true—provided the small drill hasn't wandered!

Clean up one face to be the front when finished, and set up in the 4-jaw with this one at the back so that the $\frac{1}{4}$ in. hole runs true. Start by turning off $\frac{1}{32}$ in. (say 30 thou.) and form the $\frac{15}{32}$ in. dia. shoulder. Then set over your topslide in such a way that you can taper off the web as shown. This *may* not be possible if your machine's topslide handle fouls the cross-slide in that attitude, in which case you will have to file it. Remove from the chuck and using the $\frac{15}{32}$ in. collar as a guide file round the end to rough-form the $\frac{1}{4}$ in. radius. Saw off (or file) the surplus from the sides and using a $\frac{5}{16}$ in. dia. washer with a $\frac{1}{8}$ in. hole in it on a peg in the hole rough form the $\frac{5}{32}$ in. radius at the small end. Now use finer files and finish the outline, making use of that finest measuring instrument of all (your eyes) to make all true. Round the edges and give a good rub all over with medium emery to blend all curves and straights.

Fig 6-Arch and Columns assembled.

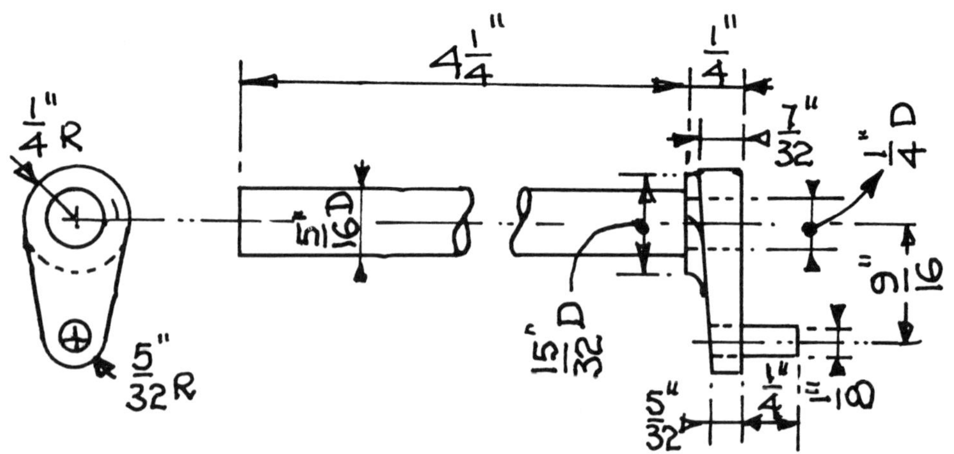

Fig 7-Crankshaft. M.S. 1 off.

18

My own crank is a tight fit aided by Loctite but I can assure you that provided the work is degreased (and primer used if you have gone all posh and used stainless steel) Loctite 601 will hold it even if you have a thou. or so slack. Finish polish, return to the lathe if need be to trim off any surplus shaft, and finally press in the $\frac{1}{8}$ in. crankpin; silver steel is ideal for this. Fig. 8 shows the finished shaft.

Flywheel (Fig. 9)

I am trying to get all the iron and steel turning done first so that you can clean down and change tooling before tackling the gunmetal—which you may prudently save these days, either to make your own castings from or to swap with Jack the Scrap for some desirable piece of metal from his yard! The casting is Stuart Turner No. 70203 and you will find it so true as scarcely to need machining. Start by going all over with an old file removing the flashes and any irregularities in the casting. I say 'old' file not so much because it's a rough job that doesn't warrant a newer one, as because a file that cuts fast may take off more metal than you mean to. Set up in the 4-jaw— see Fig. 10—gripping by the inside and adjust till true in the usual way with chalk and patience. You'll see that I am using one of the little 4 in. 4-jaws, which I find invaluable for model work as well as being (relatively) cheap.

Set in back-gear for about 40 rpm and take a cut right across the diameter with your roughing tool till you clear the surface; use power-feed, it's easier on the tool. Likewise take a skim off the front face and the back—Fig. 11 shows how I did this. You can now change to finishing tools and get a really nice polish on the machined surfaces. Use any polishing technique you like if you must, but many will prefer (on a model of this sort) a fine matt grey surface with neither toolmarks nor emery scratches showing. Now speed up to say 350 rpm and face the boss; you can machine the sides of this if you like but the usual thing was simply to turn a little bevel. Centre deeply, drill say $\frac{1}{4}$ in., and then true the bore with a little boring tool till your $\frac{5}{16}$ in. reamer enters about half an inch. (If you haven't a reamer con-

Fig 8-Crankshaft with 'blued' finish to web.

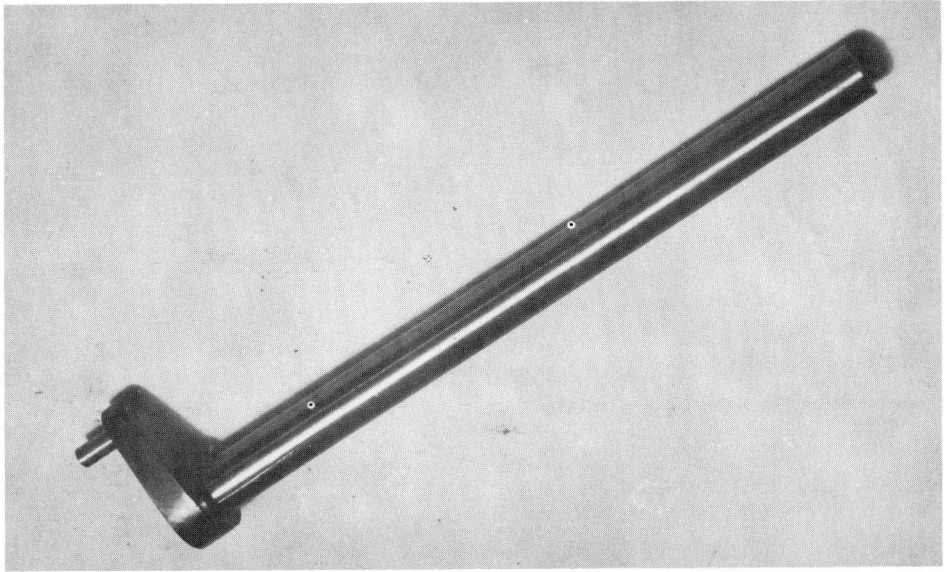

tinue boring till the hole is a nice slide fit on the shaft) I say bore before reaming because the wheel is approximately five inches in diameter and even a very slight wobble in the hole is likely to show quite badly. (These wheels usually did wobble quite a lot in the real engines, though!) (I wonder if judges take that into account at the exhibition?). Ream from the tailstock chuck at about 200 rpm, pushing the reamer once in and once out. Take from the chuck and reverse, getting the wheel as true as you can, and face the other end of the boss.

There are two grub-screw holes, at 120° in line with corresponding spokes. This is always best even when there is metal enough for larger screws than I have used. Make a good centrepop and start the hole with say a $\frac{1}{8}$ in. drill in your hand-brace otherwise there is risk of a small drill wandering off on the rough surface. I clamped the wheel to an adjustable angle plate but in default of such, screw it to a block of wood cut at about 30° to the vertical. (This does depend on how big the chuck is on your machine.) Drill both holes No. 50 or 1.8 mm and tap 8BA. The grub-screw isn't shown on the drawing. The correct procedure to adopt is to take a couple of 8BA hexagon screws, file the heads square, and then shorten them so that they grip the shaft with minimum projection of thread, but you can use ordinary grub screws if you like.

A refinement. The wheel *should* be keyed. If you have a $\frac{1}{16}$ in. square needle file you can file two keyways in the wheel and fit saddle keys. These were keys which sat on a flat on the shaft instead of the usual keyway and locked by setscrews above. They were perfectly effective on little engines like this and avoided the need for stepped keys when, as was usual, two were used on the wheel. (Stepped keys were essential, of course, as when keyways were *chipped* out it was impossible to get a perfect match between wheel and shaft on two keys; this was expensive and saddle keys were the answer.)

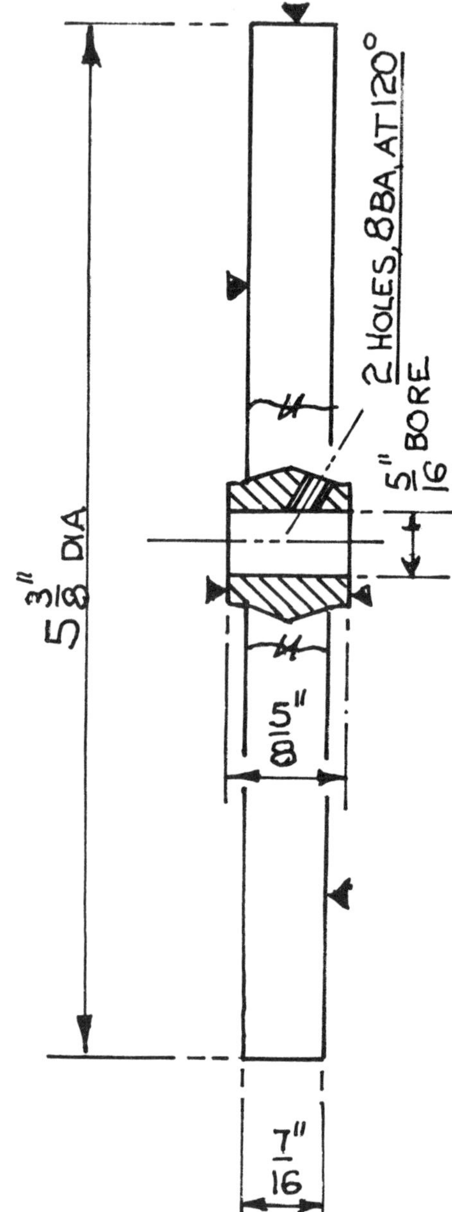

Fig 9-Flywheel. 1 off. C.I. Stuart casting No. 70203.

Fig 10-Turning the flywheel.

Fig 11-Machining the back of the flywheel rim.

Eccentric Sheave (Fig. 12) (A)

This is the last piece of cast iron—Stuart No. 60968. Chuck by the body on the 4-jaw and trim up the chucking piece. Now fit the 3-jaw and hold the casting by the chucking-piece and machine the O.D. of the sheave to $\frac{5}{8}$ in. dia., perhaps less .001 in. to give clearance when running. Grind a round nose approximately $\frac{1}{16}$ in. wide on a pointed tool and form the groove (this engages with a pip on the end of the eccentric rod)—make it about $\frac{1}{16}$ in. deep. NOTE: set this groove just a little towards the eccentric boss, to allow for machining the face later. Polish the machined surface—friction at the eccentric can be considerable and a rough sheave will soon grind away the strap.

Reset the casting in the 4-jaw this time, gripping by the chucking-piece. You should find you can just offset the casting enough to get the $\frac{5}{64}$ in. throw. If you have a DTI, use this to get the setting

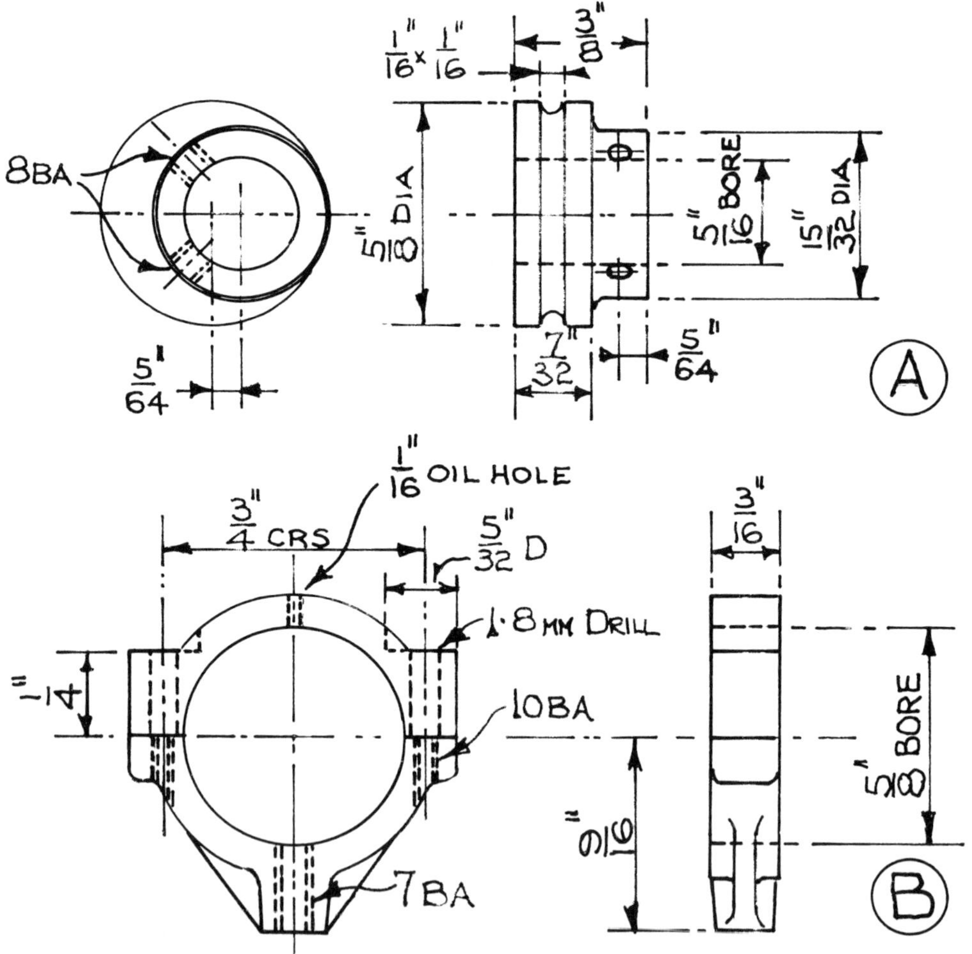

Fig 12-A. Eccentric sheave. 1 off. C.I. Stuart No. 60968.
B. Eccentric strap. 1 off. G.M. Stuart No. 60969.

right—remembering that the indicator will show the *travel* of the eccentric, which is twice the throw. If you haven't one, then use the cross-slide feed index; simply set the eccentric throw towards you, feed in till the tool just touches; rotate the job through 180°, feed in again, and the difference between the two readings is the travel $= 2 \times$ throw. Machine the boss, taking off as little as you can—it should just clean up to the $\frac{15}{32}$ in. dia.—and face the end, and the side of the sheave. Centre, and drill $\frac{1}{4}$ in. till you just reach the chucking-piece and then bore to a nice slide fit to the shaft—you want it to turn with the fingers but not by itself. Remove from the chuck, saw off the surplus and then grip by the boss to face the side. Finally, mark out for, drill, and tap the two 8BA holes; the grub-screws for this are made as for the fly-wheel. There is an added advantage in having two screws in an eccentric; when valve-setting you can put a long screw into one hole to move the sheave by, whilst the other can have a brass screw to hold it. When the setting is right (i.e. after trial under air or steam) the long screw is replaced by a proper one, which holds the eccentric whilst you change the other!

Now clean down the machine and change tooling for work in brass and gunmetal.

Eccentric Strap (Fig. 12) (B)

We might as well do this now, whilst the drawing is open at that point. The casting is in one piece and you will see it is 'oval' to give you an allowance for the sawcut. Cut it in two, and then file or mill the faces to get a flat joint. I use a fine Eclipse 'backsaw' for this sort of work, as it leaves a kerf only about $\frac{1}{64}$ in. wide. Unite the two halves (having marked them for later mating) with good quality soft solder, and get a good joint as there will be precious little metal left by the time you have finished. Set up in the 4-jaw so that the joint is central and the casting runs true at right-angles to the joint—set to the outside, not the hole.

DON'T grip too hard, as if you do the casting will give as you remove metal from the hole and then spring back, giving you a square bore! Face the outer side, and then bore to a nice fit to the sheave. Whilst still in the chuck mark out for the height of the ears and a vertical centre-line too; you can also mark out a line $\frac{13}{16}$ in. dia. to file the circular outline to later.

Take from the chuck and check that the straps still fit the sheave. If they don't, use a scraper to re-round it till it does fit. Make a stub-mandrel to fit the bore—not too tight or the solder may give way; in fact, you can make it slack, then melt a bit of resin on it and 'glue' the strap that way, easily removed afterwards. Shellac will do instead of resin. Machine the strap to the correct width and then scribe a vertical centre-line on this side also. Remove from the mandrel and file the 'ears' to be $\frac{1}{4}$ in. from the solder-joint, but don't dig into the curve of the top of the strap indicated by the $\frac{13}{16}$ in. diameter line you scribed earlier. Return to the mandrel, set the solder-joint vertical and use your scribing block to mark out the bolt-holes $\frac{3}{8}$ in. above and below the centreline.

Drilling these holes may be a problem even with a first-class drilling machine. I find the best way is to pack up the casting on the top-slide of the lathe so that its centre is exactly at centre-height and drill in the lathe; for some reason there seems to be less chance of the drill wandering and, more important, you can start the drill with a little pip from a centre-drill held in the 3-jaw. Drill clearing size for 10BA till you are *just* past the joint and then right through with 10BA tapping size.

Check that you have marked the two halves with an 00, carefully file up the outline (you may have to fit a couple of temporary screws if it comes apart) and if need be spotface the bolt-holes for the nuts. Oh-oh! I have forgotten the hole for the eccentric rod! This is done in the same way, and immediately after or before drilling the bolt-holes. The hole is tapped 7BA right through, and the boss filed or spotfaced flat to accept a locknut. As to an oil-hole, this isn't really needed, but a 1 mm hole slightly countersunk will serve. Part the two halves (if they haven't detached themselves already) lightly smooth the solder surface, remove burrs, and offer up to the sheave. If too slack, carefully file the mating faces; if too tight, get to work with a 3-cornered scraper; in any case use this to ease the bore at 20° either side of each joint, taking out about 2 thou. as the joint will tend to close up on the sheave when the bolts are tightened. This should be done on all split bearings. Fig. 13 shows the completed job.

Cylinder Set (Fig. 14 p. 26)

As mentioned earlier, this is the standard 'Williamson' set, Stuart parts Nos. 70201, 60985, 60988, 60976, but the bottom cover is made from a piece of flat brass $1\frac{1}{4}$ in. $\times \frac{3}{16}$ in. $\times 1\frac{1}{2}$ in. long. The cylinder casting must be altered by filing off the projection for the exhaust passage which, on the Williamson, is carried down through the casting, and by turning the bottom flange circular instead of the polygonal shape as cast. The drawing shows the modification at 'Section AA'. The top cover casting also is modified as we have no slide-bars on 'Georgina'.

Check the casting with a square and if need be file both the flat portface and the ends to square them up. Measure up and decide how much must come off each flange to bring the overall length to $1\frac{5}{8}$ in. and still leave both flanges more or less equal thickness; make a note of this. Set up in the 4-jaw, base outwards. It should be true to the outside of the casting and you will need packing on the side opposite the portface. Make sure the casting is square as well as true in rotation—it doesn't matter if the bore runs out a bit. Face the end to within a few thou. and then rough bore using as stiff a boring tool as you have available and for this part of the work a relatively pointed tool. When nearly to size change the toolbit (change to an even stiffer bar too, if you have one) for one with either a flat and a 45° approach angle, or with a large radiused cutting face, and finish bore. Especially if your toolbits are carbon steel it is better to use a relatively coarse feed and a wide cutting face than to use the conventional 'very fine feed'. With the latter the tool has to run so many miles per cut that it is abraded away before it gets to the end and you have a taper bore. I used up to 8 thou/rev and a tool with a flat face in a bar almost $\frac{1}{2}$ in. dia. and had no chatter, but you can always reduce the speed if that develops. The final cut should be at least 2—3 thou. deep—the tool must *cut*, not rub, and this means it must be honed dead sharp before the final cut. If this leaves the bore a few thou. up or down, not to worry, as you can make the piston to fit. Incidentally, if you have a relatively flexible bar and suffer chatter, try parking a piece of

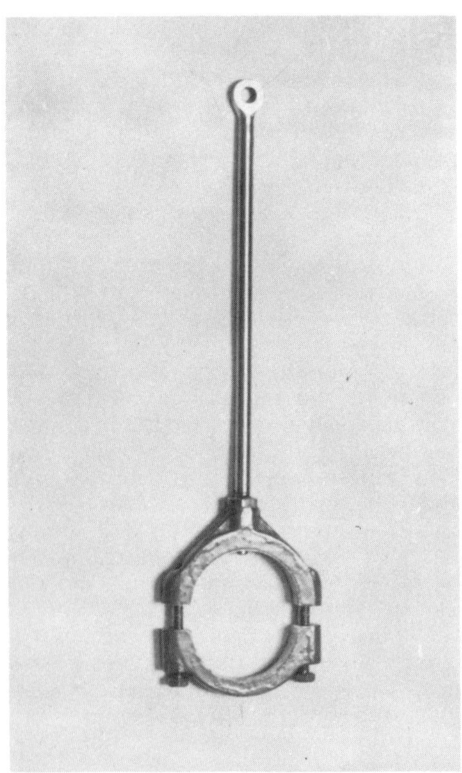

Fig 13-The completed eccentric straps and rod

Plasticine on the end, to damp out the vibrations.

You can now turn the outside of the flange and the cut can be taken right up to the bottom of the portface projection. The shape can be trimmed off later to give a good appearance. Finish face the flange. Mark out centrelines whilst in the chuck, squaring up to the portface, and whilst you are at it, mark out also for the $\frac{5}{8}$ in. dimension on this face. See Fig. 15. In this picture the bottom flange, as you will see is not yet turned. Set the work on a mandrel—it's easier if you make a stub-mandrel held in the chuck—and face the other end to bring the overall length to $1\frac{5}{8}$ in. and with a sharp tool scribe round a circle $1\frac{1}{8}$ in. dia. to act as a guide when filing up the top outline. Furthermore I suggest you relieve the top face of the port block about 5–10 thou., too. Finally, pick up the already scribed centrelines, etc. and carry these round onto the parts of the casting which you were not able to reach before.

Remove from the lathe and set up the vertical slide with some care as to squareness. Mount the cylinder as shown in Fig. 16 and fit the sharpest milling cutter you have in the 3-jaw. Now, this may not run quite true, so it's worth while experimenting with paper packing under one or more jaws till the cutter runs with no wobble. Set the cylinder square to the marks (don't forget the washer, preferably brass, under the nut) and engage the leadscrew handwheel to apply the cut. Machine the whole down to the line. Whilst this set-up is available you can also drill the ports. Fig. 17 shows the 'co-ordinates' of the holes; set a small centre-drill in the chuck and offer this to the top of the cylinder adjusting the cross-slide till it is exactly in line. This is your zero, so either adjust the index dial (if a Super-7) or make a note of the setting if it's a fixed one. Similarly adjust the vertical slide so that the point of the drill is on the centreline. Make a note of the setting. Now, raise the vertical slide 0.063 in., wind in 0.375 in., and make a pop with the centre-drill. Wind in a further 0.375 in. (0.750 in all) and repeat. Wind down the vertical slide by 0.126 and make another; cross-slide back by 0.375 and repeat. Finally, return the vertical slide to the zero or reference position (and, I suggest, the cross-slide also) and feed the cross-slide forward 0.563 from zero to pop the exhaust port. Now repeat the whole operation, using the correct sized drills, and using your leadscrew handwheel to feed in the proper depth each time. (The exhaust hole should be the same depth as the others; use a very sharp drill for this one, as a blunt drill may exert enough pressure to binge in the cylinder bore—the metal is pretty ductile') Don't remove the vertical slide.

You now need to cut the two recesses at each end of the cylinder which pass the steam from the passages to the bore. Reclamp to the vertical slide with the cylinder ends facing the headstock and the bore parallel to the lathe bed. Adjust the slide till the cylinder centreline is at lathe centreheight and put a $\frac{1}{4}$ in. dia. endmill or, better, a slot-drill in the chuck. Advance the saddle till the cutter just brushes the surface and make a note of the leadscrew handwheel reading. Retract the cross-slide so that the endmill can enter the bore and put on about $\frac{1}{16}$ in. of cut, traverse across till you have fed in just over $\frac{1}{8}$ in., and then repeat with a further $\frac{1}{32}$ in. of cut to go full depth. If your cutter is a slot-drill of course you can plunge straight down and do the job in one go. Repeat for the other end, and then remove burrs.

To drill the passages—easy if you have an adjustable angle plate or vice; simply set to the angles shown and drill. However, do set your depth stop, as it is all too easy to go too deep. If you have no such luxuries, then use a block of wood cut as near as may be to the right—correct, that is—angle as shown in Fig. 18. This is a simple method and after a few years you will have a collection of such little wedges and, with these, who needs to spend money on angle-vices? The woodscrew shown is quite adequate

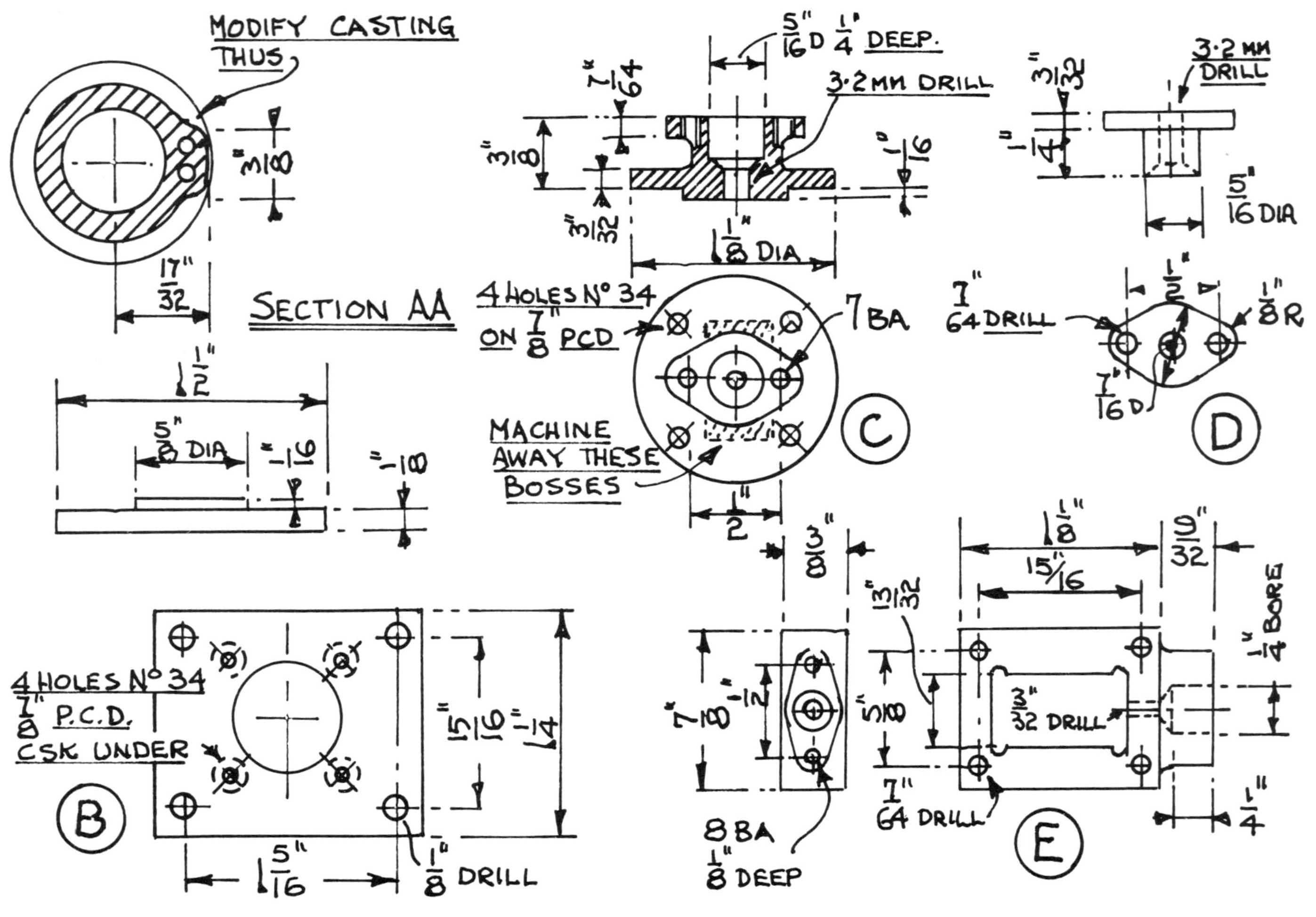

MODIFY CASTING THUS
3/10"
17/32"
SECTION AA
5" D 1/4 DEEP. (5/16 D 1/4 DEEP.)
3.2 MM DRILL
7/64"
3/8"
3/32"
1/16"
1 1/8 DIA
3/32"
1/4"
3.2 MM DRILL
5/16 DIA
4 HOLES N° 34 ON 7/8 PCD
7 BA
7/64 DRILL
1/2"
1/8 R
7/16 D
D
C
MACHINE AWAY THESE BOSSES
2"
1/2"
5/8 DIA
1/16"
1/10"
4 HOLES N° 34 7/8 P.C.D. CSK UNDER
B
15/16"
1 1/4"
5/16"
1/8 DRILL
3/10"
7/8"
1/2"
5/10"
8 BA 1/8 DEEP
13/32"
1 1/8"
15/16"
9/32"
3/32 DRILL
1/4 BORE
7/64 DRILL
1/4"
E

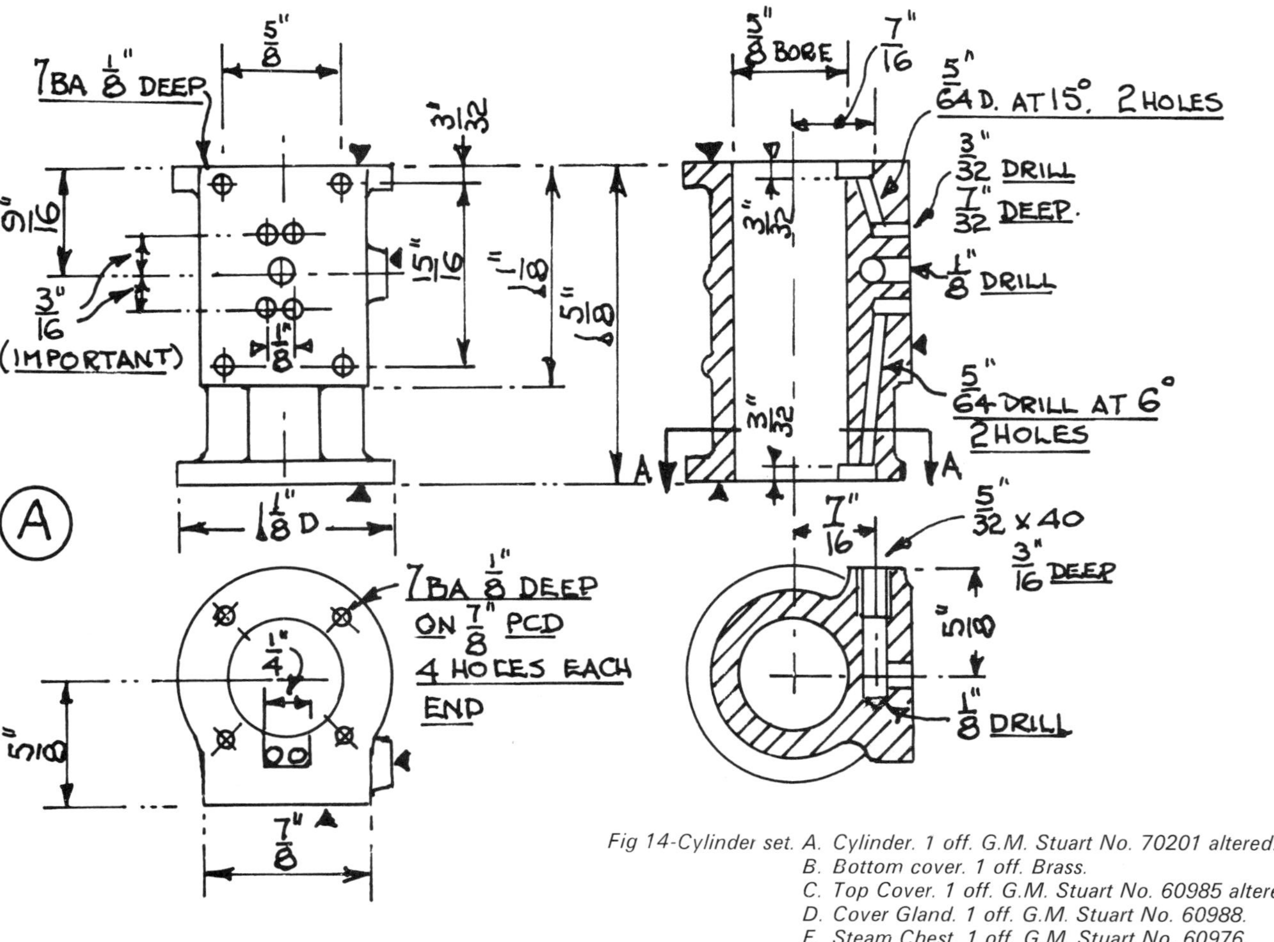

Fig 14-Cylinder set. A. Cylinder. 1 off. G.M. Stuart No. 70201 altered.
B. Bottom cover. 1 off. Brass.
C. Top Cover. 1 off. G.M. Stuart No. 60985 altered.
D. Cover Gland. 1 off. G.M. Stuart No. 60988.
E. Steam Chest. 1 off. G.M. Stuart No. 60976.

Fig 15-How to mark out cylinder centres.

Fig 16-Set-up for machining the portface.

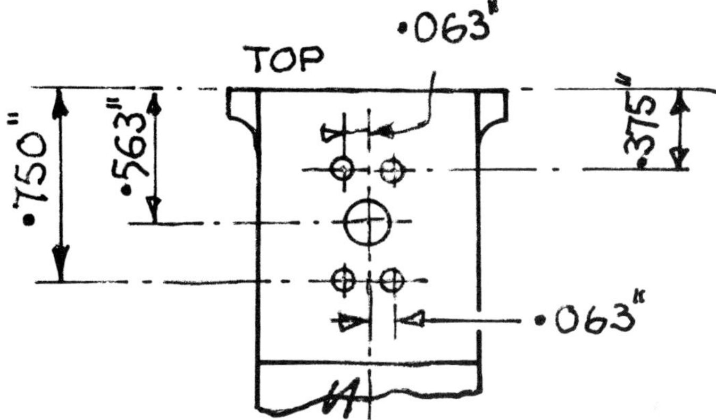

Fig 17-Co-ordinates for the steam and exhaust ports.

holding for drilling these holes or even larger ones.

The final hole to be drilled at this stage is that for the 'entrance to the way out' for the exhaust. This needs little comment except to take care that the job is set true when drilling and to be cautious just as the drill breaks through. Tap the end $\frac{5}{32}$ in. × 40 tpi (or to please yourself) for the exhaust elbow. This having been done, go over the casting trimming up any roughness, removing all burrs and— important—put a 30° bevel at the ends of the bore with a scraper to ease entry of the piston packing.

Bottom Cover (Fig. 14) (B)

Select a piece of brass flat which *is* flat, or choose the concave side for the bottom. File it rectangular using care and a try-square and then set it on packing in the 4-jaw so that it does lie flat. With a knife tool chew off the face leaving a boss $\frac{1}{16}$ in. proud and just over $\frac{5}{8}$ in. dia. Then finish turn, making this projecting boss a nice fit to the cylinder. Mark out for the holes. Now, four of these are shown at $\frac{7}{8}$ in. PCD, and surprise, surprise, this works out that they are within a thou. or so of $\frac{5}{8}$ in. apart, so you can mark them out with a scribing block, just like the others. Remove from the chuck and mike round the $\frac{1}{8}$ in. thickness. If this is more than the odd thou. out, get busy with a file and bring it to this tolerance,

otherwise the cylinder won't sit up straight. Apart from looks, this plays havoc with the guiding of the piston-rod. Countersink the four $\frac{7}{64}$ in. holes on the underside.

Top Cover (Fig. 14) (C)

This is Stuart part number 60985. First, saw off the surplus metal as shown and trim round the casting removing flashes etc. Grip by the chucking-piece in the 4-jaw and let it project just enough to be able to machine the O.D. of the flange. Face the gland-boss, centre, and drill $\frac{17}{64}$ in., followed by a little boring tool till it is $\frac{5}{16}$ in. dia. Recentre with a small slocumbe and drill through 3.2 mm, which will give you just enough clearance on the rod. With a parting tool turn away the metal behind the gland-flange to leave this $\frac{7}{64}$ in. thick, and by repeated applications bring the parallel part back as far as the face of the cover. You should then be able just to get a side-cutting tool in to face the cover proper. Mark out for the position of the cover bolts, (same as for the bottom cover) first having drawn a centreline across the long way of the gland oval as a reference line—you want the oval to stand at 45° to the fixing studs. Remove from the chuck, fit the 3-jaw, and make a stub mandrel to fit the gland hole. You may need a 6BA holding screw, but the shellac or resin glue will serve as well. Saw off the

Fig 18-Drilling steam ports with a 'rare and costly' substitute for an angle-vice.

chucking piece and machine the under face to the dimensions shown, the spigot a nice fit to the cylinder.

The *gland* (D) (S.T. Part 60988) is a simple turning job, the body being made a slide fit in the hole in the cover. Mark out in the chuck for the shape of the oval and for the bolt-holes. A detail not shown on the drawing—lightly countersink the 3.2 mm hole on the outside, to act as an oil sink. You can now have a rest from the lathe and proceed to the drilling machine. Drill the holes in the gland and then use this as a template to drill the corresponding tapped holes in the cover. This done, you may care to trim up the ovals so that they look right; they aren't the same size on cover and gland but they should look a pair if you follow me.

Drill the holes in the covers and offer these to the cylinder. Do the bottom one first; fit the cover and lay the assembly on its side on the steam-port face. This will enable you to get the cover square and after clamping, spot through $\frac{7}{64}$ in. and then drill 7BA tapping size. Attach this cover, and put on the top one. The bottom cover will then give you a line to eye from to get the gland studs in line with the long centreline of the bottom cover. Spot through one hole, drill and tap, drop in a temporary bolt and see how it looks. If need be, draw the hole with a file till the cover sits square. Then tighten this screw and spot through the other three holes; drill and tap 7BA.

Steam Chest (Fig. 14) (E)

The chest is Stuart part 60976. Grip in the 4-jaw to face both sides to $\frac{3}{8}$ in. thick (or just file them) and then use a square to scribe lines round the rectangular shape at right angles, together with a centre-line. File to these lines to true up the outline— leave a 'rough emery' finish to act as a key to paint. Now set in the 4-jaw with shims and thicker packing to avoid damage to the faces to machine the gland-hole (Fig. 19). Then face the gland boss, and drill as already described for the similar hole in the top cylinder cover. Mark a centreline across the gland whilst in the chuck.

The insides of the chest are a little different from the usual rectangular box, and form a guide for the slide valve. There is rather a lot of metal to come out, but it is only necessary to use a file taking some care to keep the $\frac{15}{32}$ in. width central, and parallel to the valve travel. The valve itself is a brass pressing, almost spot on to size, and after lightly trimming the sides can be used as a gauge. It will be necessary to file off between 10 thou. and $\frac{1}{64}$ in. at each *end* of the chest cavity to give enough room for valve travel. Now mark out for the four fixing holes and drill them. Use these as a template to drill the cylinder, clamping the chest to the cylinder, as it is important that it does not move when spotting through.

Fig 19-Use of packing when holding steam chest in the 4-jaw chuck.

Chest Gland (Fig. 20) (A)

Casting is S.T. No. 60981. The easiest way to make this is first to grip by the gland body and turn the projecting rod guide to say $\frac{7}{32}$ in. dia. parallel. Then reverse in the 3-jaw and machine the body to a nice fit to the chest, centre and drill $\frac{3}{32}$ in. (it's not worth reaming, but after assembly put a $\frac{3}{32}$ in. drill through both gland and chest hole in case there is any misalignment) and finally, offer a $\frac{1}{4}$ in. drill to form the nominally 120° recess. Reverse, and grip this part in the 3-jaw to form the taper shape of the rod guide projection. NOTE: don't tighten the chuck too much; there must be hundreds of engines about with slightly triangular glands due to overtightening at this stage!

Chest Cover (Fig. 20) (B)

S.T. part No. 60978. This presents very little work—All that is necessary is simply to file the bedding face, then drill the holes and leave it at that. You will, I believe, find the steam entry boss to be quite a little taller than is illustrated but that is easily rectified. Whether you file or spot-face this is up to you. The steam joint will be made on the threads unless you are going to use an unsightly union at this point and on my engine the whole is painted. It is prudent to get all four stud bosses the same height, or you will have odd lengths of thread projecting beyond the nuts.

Piston and Rod (Fig. 21)

The piston should be made from hard brass. Chuck a piece of $\frac{3}{4}$ in. stuff in the 3-jaw, about $\frac{7}{16}$ in. protruding. Face the end, and then turn the body to about 5 thou. *over* the correct diameter—the final bringing to size is done after attaching to the rod. Drill No. 37 and tap 5BA. To get the hole straight, fit the tap-wrench part way down the shank of the tap and guide the tail end with your tailstock drill-chuck, lightly gripping the shank so that it guides but doesn't grip, if you follow me. Take the seconds tap in just over full depth—a slight tightness on the rod will be an advantage here. Now, with your parting tool form the recess for the packing; the dimensions are not critical, but don't make it any

Fig 20-A. Valve-rod gland. 1 off. G.M. Stuart No. 60981.
B. Valve Chest Cover. 1 off. G.M. Stuart No. 60978.

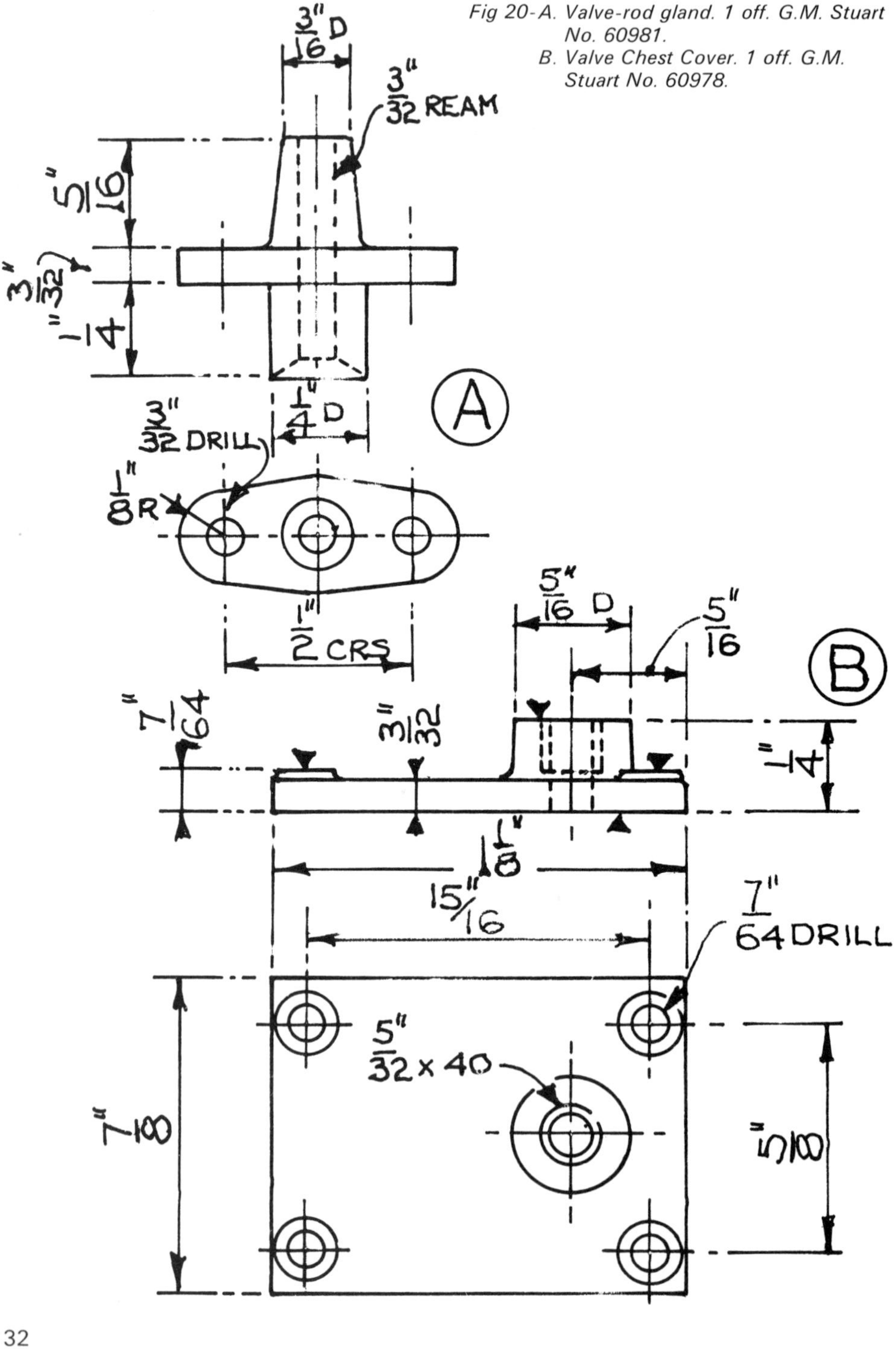

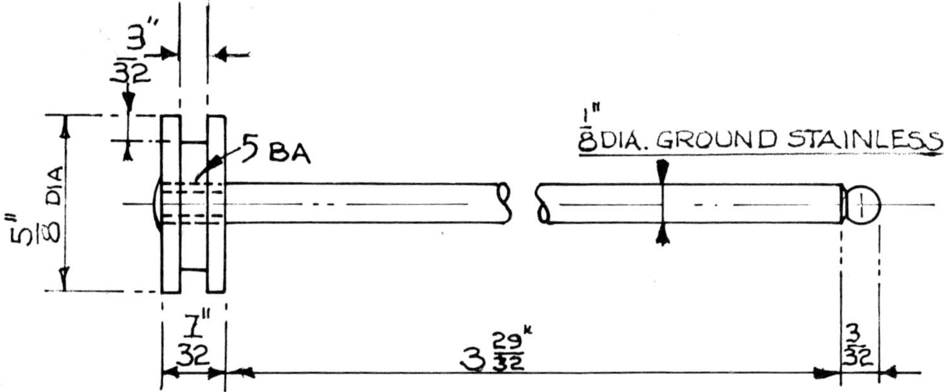

Fig 21-Piston and Rod. 1 off. Brass with Stainless Steel rod.

shallower—deeper is better. Part off a shade over thickness.

The rod should be ground stainless and, as you can now get the free-cutting variety from most advertisers, presents none of the problems that once beset screwing such material. (I always screw-cut the old stuff and finished only with the die—still do on important jobs.) Provided your 3-jaw is reasonably true there is no need for special steps at this stage. First, cut the rod a shade over length (say $4\frac{1}{4}$in.) put in the chuck and make the little knob—this is purely decorative; for some reason Victorian engineers liked knobs on the end of such rods. Reverse and use your tailstock die-holder to cut the thread 5BA. Set the die first on a well-made stud or bolt and try to cut in one go. Form a 45° bevel on the end of the rod first, but do it with a tool not a file or the bevel may throw the thread drunken. Remove the burrs at the ends of the thread and polish the rod.

You must now attach piston to rod, firmly. The best way is to chuck the piston, apply a little Loctite, grip the rod in the tailstock chuck, and pull the 3-jaw round by the chuck key till the rod is well home. Remove, and rechuck the rod with enough projecting to get a Dial Indicator to it. Fit paper under one jaw or another till it runs true (or use a copper shim) and then push the piston back till there is just enough room for a tool to clear the chuck jaws. Turn the O.D. of the piston till it is between .001 in. and .003 in. clear in the cylinder—no more. Turn off any surplus rod which may project. There is no need to true up the faces or the groove—if the thing wobbles so much that you feel you must, you are unfortunately faced with having to make a new piston, or a new rod. NOTE: Some people like to rivet the rod to the piston. I have no objection, but it does make it difficult to take apart if that ever becomes necessary.

Valve and Rods (Fig. 22)

The valve is Stuart No. SX138, and the eccentric rod No. 60971. The slide valve needs little save smoothing the face and adjusting the slots on the back. The steam edges may need touching with a very fine file, but that's all—I think your micrometer will show that it is almost spot on $\frac{1}{2}$in. overall. Smooth the face on a good, fine, flat file and then with finest emery sheet, new, held dead flat on say the lathe-bed. Finish with the scratches running across the valve. Now, with marking blue, check that the valve sits flat on the port face, and adjust as required. I find it almost always necessary to do a bit of work with a scraper (a very sharp one) to get good contact, but don't be misled by getting either too much or too little blue on the face. The cross-slot will have to be deepened a bit, but I suggest you leave this till you assemble; don't forget to do it, though, using a square needle-file. The width of the slot is usually just right.

The *valve nut* (D) can be made from $\frac{1}{8}$ in. × $\frac{3}{16}$ in. in brass flat, if you find this fits snugly in the slot. If you haven't any, make it in the lathe. Chuck a piece of $\frac{3}{8}$ in. round brass, drill and tap 7BA, and part off $\frac{1}{8}$ in. thick. Then file to shape. NOTE: The 7BA hole is shown in the middle of the $\frac{3}{16}$ in. width, but it can with advantage be say $\frac{1}{64}$ in. offset; this reduces the need to deepen the valve slot. The *valverod* (C) needs little comment, being no more than an exercise in using the tailstock die-holder. However, you WILL find after screwing that the thread won't go through the gland. File off the tops of the longer length of thread till it will. The *valve-rod head* (B) is made from $\frac{3}{16}$ in. square steel, though no reason why you shouldn't use brass. Chuck in the 4-jaw truly central (don't you wish you had a self-centring *four*-jaw chuck? Worth thinking on if your Premium Bond comes up; I got one second hand years ago, and it's one of the best investments I ever made). Turn the square end round for $\frac{1}{8}$ in. long, as shown on the drawing, centre, drill 7BA tapping size about $\frac{1}{2}$ in. deep and tap—it's easier to do this in the lathe than in the vice. Part off a shade over $\frac{13}{32}$ in. long. Mark out for and drill the cross-hole, noting that it is 2.5 mm one side and 7BA the other, so drill 7BA tapping first and then open out. Taking care to get the slot square saw down for the $\frac{3}{32}$ in. wide slot and widen it with a fine warding file. Tap the one side of the hole and finally form the rounded end; clean up, remove burrs, and mildly polish.

The eccentric rod (Fig. 22)(E), comes partly made, Stuart No. 60971. The eye is ready welded to the rod, but will need trimming to width and the rod will need shortening. Cut off the surplus and then in the 3-jaw form a new $\frac{1}{16}$ in. pip on the end, to fit the groove in the eccentric. Use the tailstock dieholder to extend the thread to $\frac{5}{16}$ in. long. Finally, fit the eye end to the slot in the valve-rod head. The drawing doesn't show the pivot pin—this is no more than a 7BA bolt with a plain

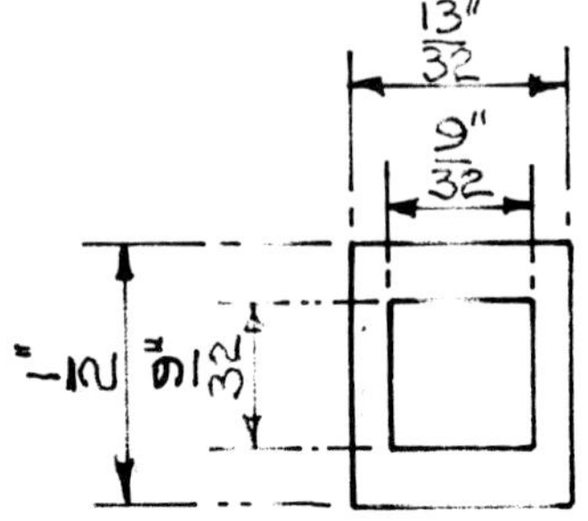

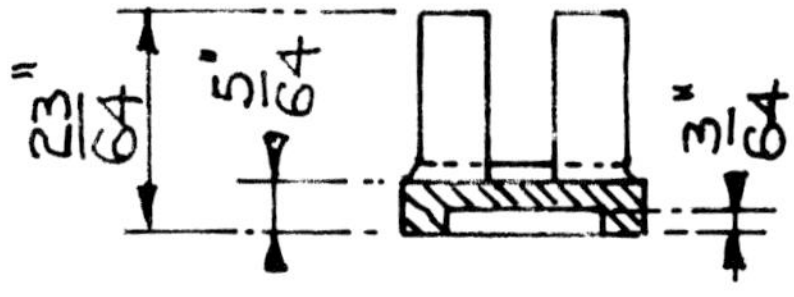

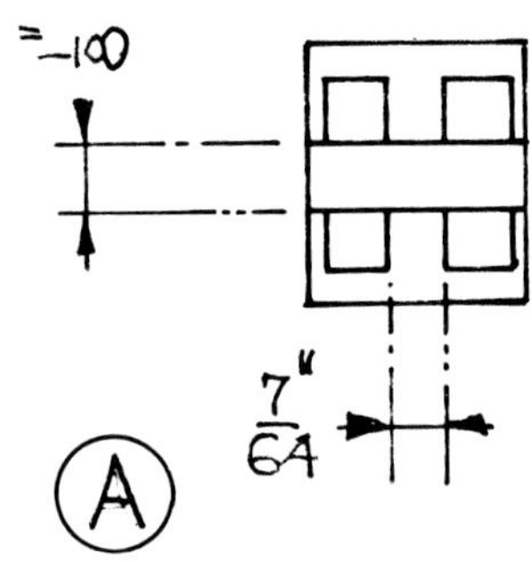

Fig 22 A. Valve. 1 off. Hot pressed Brass. Stuart No. SX138.

shank, shortened to $\frac{1}{4}$ in. overall length and threaded to $\frac{1}{8}$ in. One point, the eye of the rod may be slack on the bolt. If it is, turn a little brass peg, drive this in, and redrill 2.5 mm.

Now we come to the really interesting parts. So far we have been using workshop processes which most of you will have used before but now it's all handwork and in one sense real craftsmanship. I think you'll enjoy it, even if you do

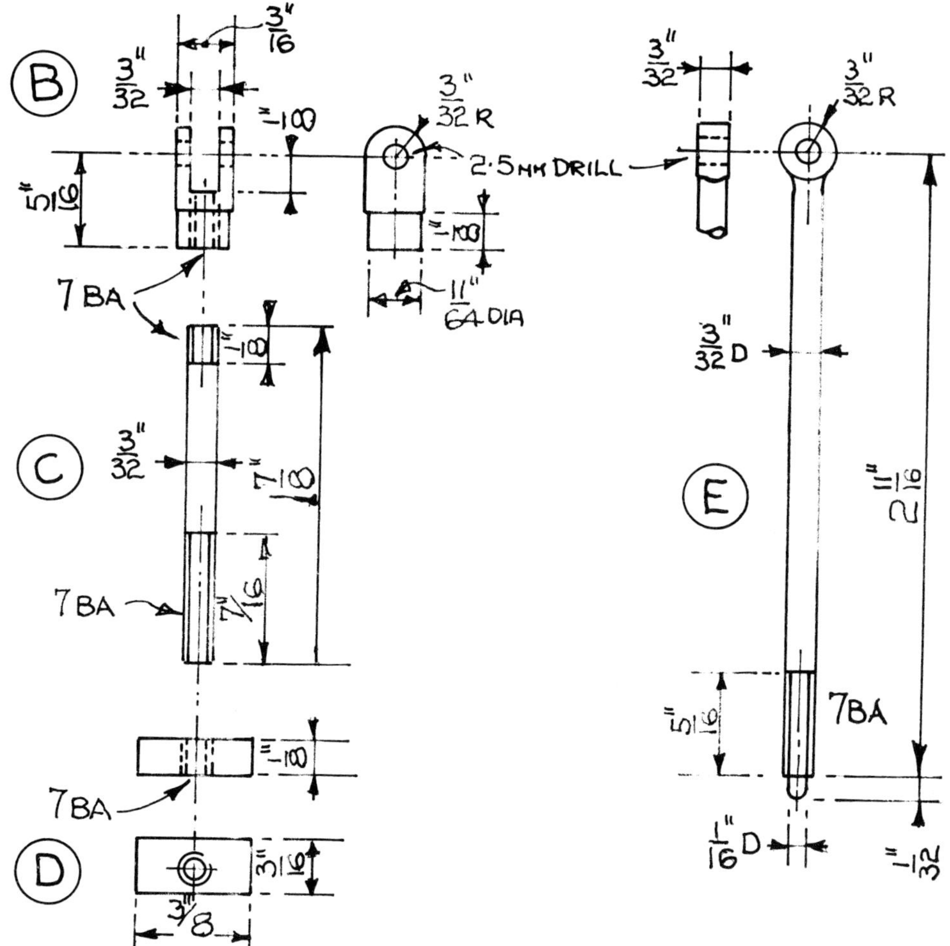

Fig 22-B. Valve Rod Head. 1 off. M.S.
* C. Valve Spindle. 1 off. Stainless Steel.*
*   (Or bronze).*

D. Valve Nut. 1 off. Brass.
E. Eccentric Rod. M.S. Finished Part
Stuart No. 60971 altered.

make a few boss-shots at first. You are going to need a small—say $\frac{1}{4}$lb—ball-pein hammer, though a cross-pein tackhammer may serve, and an anvil of some sort. My vice is one of the so-called 'Garage Vices' which has an anvil on the back, but a piece of $\frac{1}{2}$in. wide steel flat in the vice jaws will serve quite well. Round the sharp edges and polish the top with emery cloth. You also need a ready source of heat and something more like a hefty bunsen burner or the larger type of primus stove is best. In short, you are going to try your hand at blacksmithing.

Piston Rod Guide Bridle (Fig. 23)

The photo, Fig. 24, shows the finished article, and you can see it also on the picture of the engine shown at the start. It is made of mild steel, though if you have a scrap of wrought iron this is better. I started with a piece of $\frac{3}{8}$in. $\times \frac{1}{8}$in.

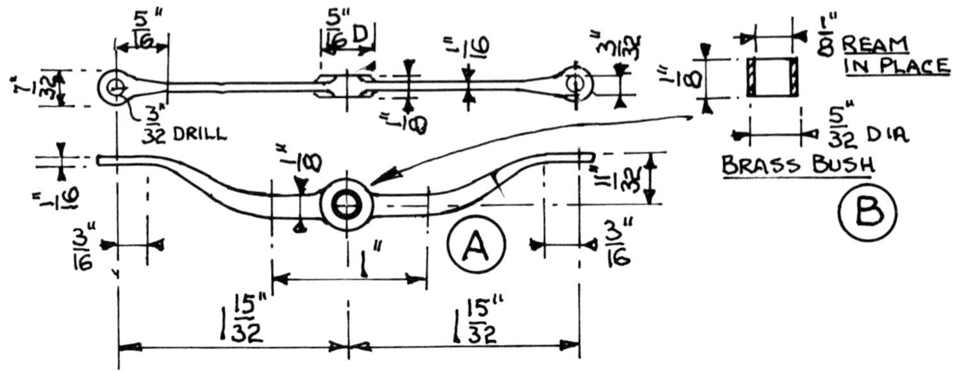

Fig 23-A. Bridle. 1 off. Forged M.S.
    B. Bush. 1 off. Brass or G.M.

Fig 24-Finished Bridle. This specimen was made entirely by forging with the hammer.

about 3 in. long. Put a light centre-pop in the very middle and draw a circle $\frac{5}{16}$ in. dia. Tangential to this circle file grooves about $\frac{1}{8}$ in. wide, both sides of the circle and both sides of the flat, just over $\frac{1}{64}$ in. deep. These are really a pretty hefty form of marking out. Now get one end hot—really hot—and as quickly as you can transfer to the anvil and rest the work thereon with the groove just aligned to the edge. Swage the material down, turning over and hitting both sides if the metal stays hot enough long enough. Repeat till you have reduced the thickness to $\frac{1}{16}$ in. Note that the material will tend to widen as well as lengthen, so occasionally give light blows on the sides to reduce this. Do both ends, and *don't be in a hurry*; it's so small that the only way is to keep heating and get as many light blows in as you can. If you

hit too hard and dent it below $\frac{1}{16}$ in. thick you must start all over again.

It will probably curve and go crooked, but don't worry—just heat up and straighten it. Don't do anything cold, always heat.

You will now have a piece which is more or less $\frac{3}{8}$ in. wide, a square boss in the middle and two arms $\frac{1}{16}$ in. thick. Now get out your small files. Form the centre boss to circular, and file down the width to the shape shown in Fig. 25. Once filed more or less to shape, heat up and give a few token blows with the hammer; just tap it, to get the 'as forged' patina. The centres of the end bosses should be about $3\frac{1}{8}$ in. apart at this stage. At this point I drilled the holes, but I think the end ones could perhaps be drilled after the next stage—I didn't try. This next stage is to get the end hot and

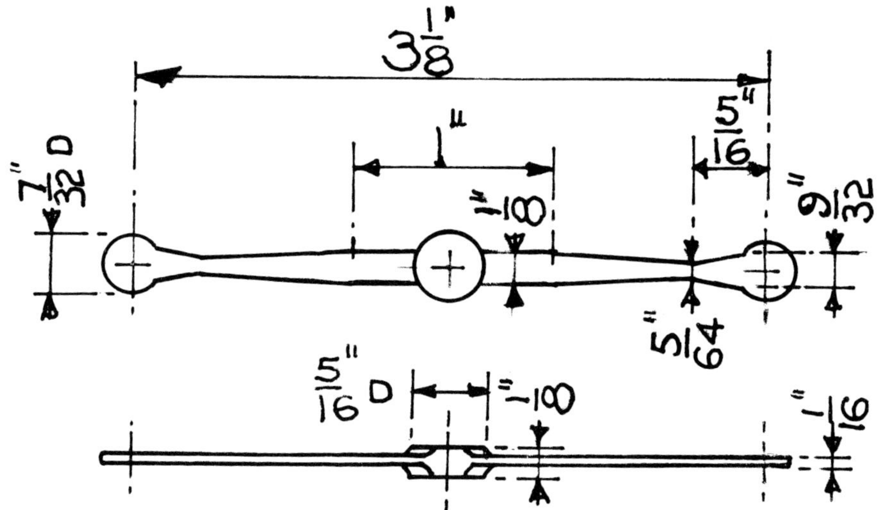

Fig 25-The second stage in making the bridle.

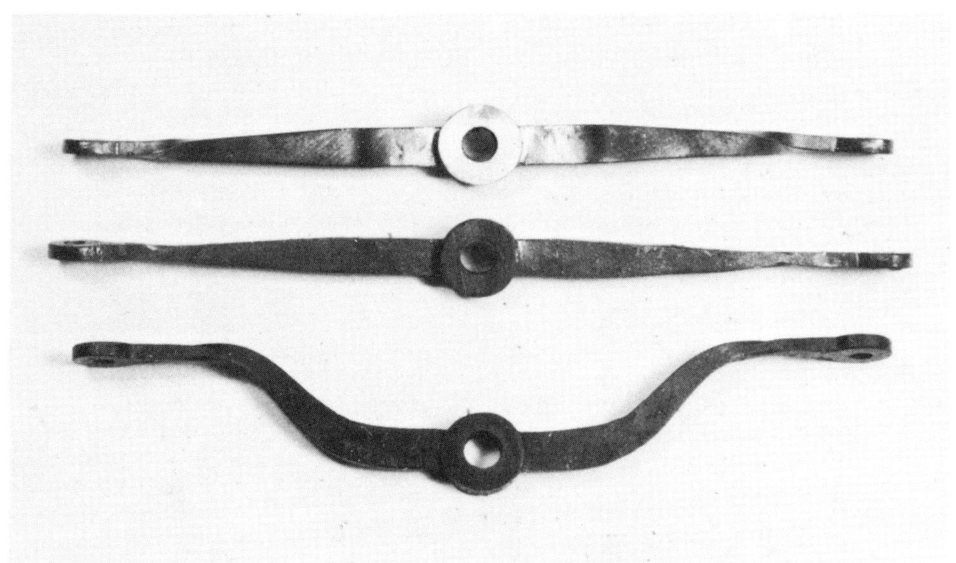

Fig 26- Steps in making the Bridle.
Top. After twisting the ends.
Centre. The ends hot-hammered. A little more work would have improved the appearance.
Bottom. Formed to shape.

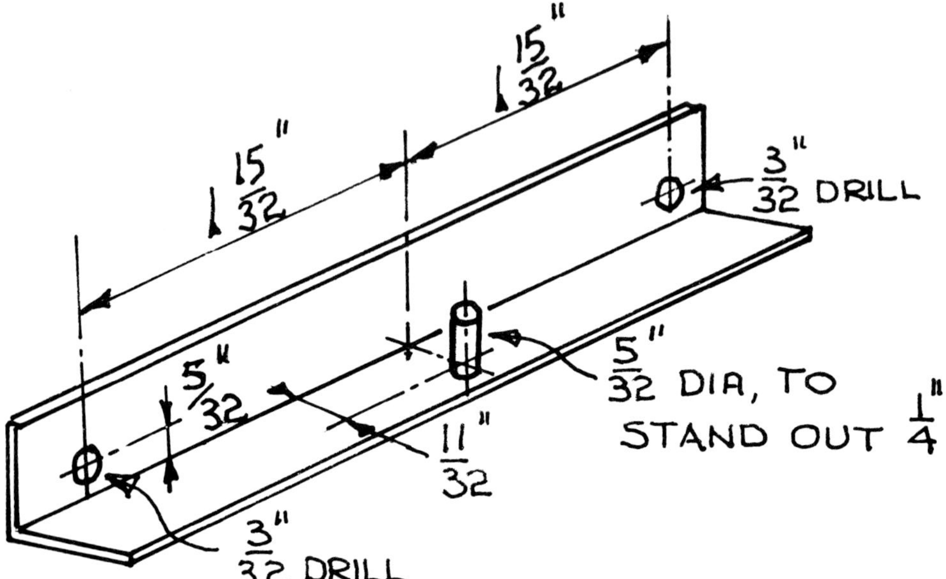

Fig 27-Jig used in forming the bridle.

twist, taking care that each end is twisted in the opposite way, so that the same face of the bar when in the flat faces the same way when twisted. Fig. 26 shows this; the top view just after twisting, the middle one after giving a few blows of the hammer on the (heated) twist to even out the material. It shouldn't look like a twisted candlestick.

The lowest view in Fig. 26 shows the final 'cupid's bow' bend (compare with Fig. 24). To achieve this needs a bit of jiggery. Find a piece of steel or brass angle, about $\frac{1}{2}$ in. × $\frac{1}{2}$ in. or so and say $3\frac{1}{2}$ in. long. Mark out the centreline of the length on the inside of the angle, both flanges. In the bottom, drill a hole to accept a peg to fit the centre-hole of the bridle ($\frac{5}{32}$ in.) exactly $\frac{11}{32}$ in. from the other flange. In this latter, drill two holes at $1\frac{15}{32}$ in. from the centreline, both at the same height above the bottom flange; these to pass 8BA—say $\frac{3}{32}$ in. (see Fig. 27). Note that if you have adjusted the centres of the engine columns to suit a slightly non-standard arch, then adjust these dimensions also.

Put the bends in the wide part of the bridle first, and then at the two ends.

Offer up to your little jig and adjust till when slipped over the centre peg an 8BA bolt can pass through the holes in both jig and bridle. This will take a bit of fiddling, and you may find it ends up *like* a cupid's bow, but that doesn't matter so long as it's not too much so. The final check is to remove the centre peg, bolt up to the other two holes, and fit a $\frac{5}{32}$ in. mandrel (a bit of silver steel) in the centre hole and check that this stands upright. If not, just strain it a bit this way or that till it is. The final job (purely 'cosmetic') is to get all red-hot again, allow to cool till just barely visible red, and quench in old engine-oil. This will give a more or less permanent 'blacksmith' effect. But mark one side of the centre boss with an 'O' first—this will be the top when installed in the engine. As I said at first, you may have to make a couple before you are satisfied with the job—I made three before I sorted out how to do it—but it's interesting work and the result worth showing to friends. The main snag, of course, is that the metal doesn't keep hot very long and even a time-served blacksmith would find it difficult to forge completely. One point—

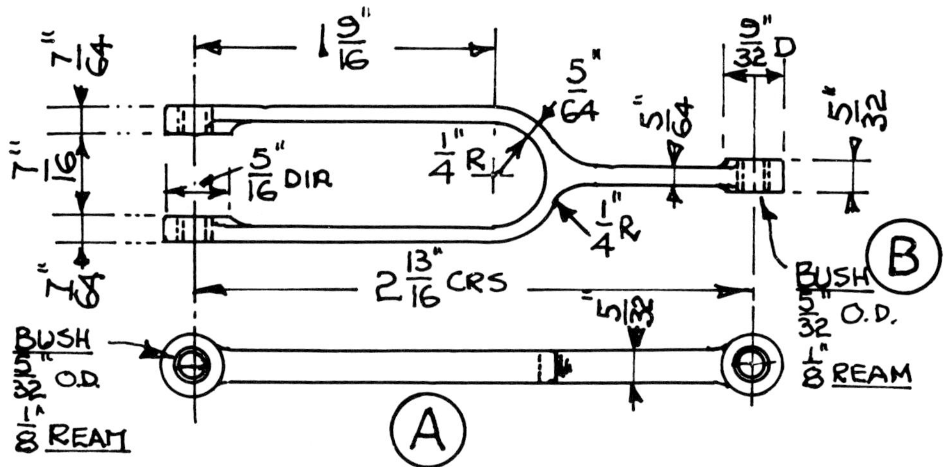

*Fig 28-A. Connecting Rod. 1 off. M.S. fabrication.
B. Bush. 3 off. Brass or G.M.*

keep the widths and thicknesses heavy rather than light. It's a bit fragile otherwise.

The bush is best made by drilling No. 31, pressing the bush in place, and opening out to 3.2 mm afterwards. The drawing says 'ream' but you will almost certainly find this too tight a fit. However, as it's easier to open out a hole than to shrink it try $\frac{1}{8}$ in. reamed first.

Connecting Rod (Fig. 28)

This CAN be forged throughout—I made one that way—but it's a terrible job to do, and to be quite honest the final result did not look all that good either; the main snag being that the holes couldn't be put quite in the middle of the bosses due to forging errors. So, there are three ways of doing it. One, which I don't propose to describe, is to take a piece of $\frac{3}{4}$ in. $\times \frac{5}{16}$ in. steel flat about 3 in. long and literally carve the thing from the solid. This is just a case of successive marking out, drilling, sawing, and filing. Fair enough if you like that sort of thing, but I don't! However, the end result will be effective and if you find the following description too formidable, do it that way. *But* before starting get the piece red-hot and allow to cool slowly, to relieve the internal stresses which are ever-present in bright-drawn steel.

The two methods I recommend are similar; in the first, you bend a piece of strip to form the U and braze on the lower bosses and upper stalk complete with boss. Whilst in the second a piece of strip is first split part way down, the U formed from these two legs and the upper stalk from the unsplit part. All three bosses are then brazed on. The first is easier, the second gives the blacksmith's look without the bright brazed joint in the middle. To deal with these in turn.

Take a piece of $\frac{3}{16}$ in. $\times \frac{3}{32}$ in. strip say 4 in. long and hot-forge it down to $\frac{5}{64}$ in. $\times \frac{5}{32}$ in. This is easy; just hold it in your mole-grip or a pair of pliers and work first on one end then the other. Now grip a piece of $\frac{1}{2}$ in. round bar (a little smaller is better) in the vice, tightly, heat the centre of the strip and bend it round the bar. You will have to use the hammer a little to get the bend close. Check the distance between the legs and do any straightening necessary. Make the bottom bosses in one, $\frac{11}{16}$ in. long, complete with the $\frac{5}{32}$ in. hole, and file a flat on each end so that it is $\frac{1}{8}$ in. from the centre line of the boss (Fig. 29). Cut off the legs of the 'U' square across so that they are $1\frac{11}{16}$ in. from the inside of the 'U'. Then, when the boss is brazed on it will

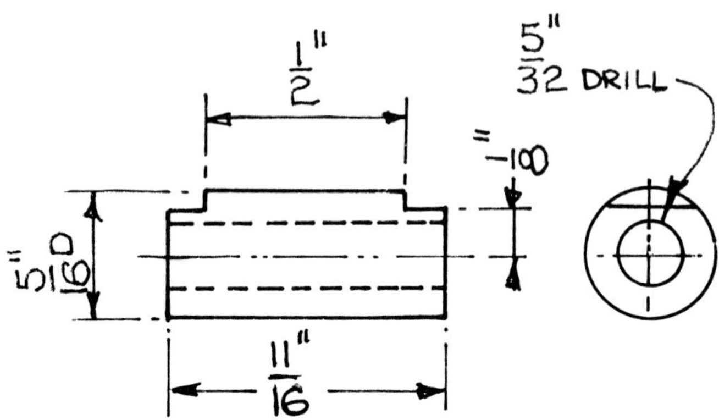

Fig 29-Crosshead end of connecting rod before brazing in place.

meet the $1\frac{9}{16}$ in. dimension shown. Don't braze this on yet.

To make the leg, forge down (or file it if you prefer) a piece the right shape from $\frac{3}{16}$ in. $\times \frac{5}{16}$ in. steel. Drill the boss $\frac{5}{32}$ in. and cut off to length so that the centre-distance between the hole in this leg and those in the 'U' will be $2\frac{3}{16}$ in. File off the scale from the U locally for the brazed joints and from the leg—just clear enough off to make the joint and a little fillet; flux won't clear the scale and this acts as a resist to prevent too much silver solder spreading about. Assemble the parts, fluxing well, and hold in place with two rods and wire as shown in the photo, Fig. 31. (This is actually the other method, but the same device is used.) Braze up the joint between leg and U first, then transfer the torch to the two lower joints and braze these. Check that the thing is square—both rods parallel— before doing this last joint and if anything has shifted correct it first. Pickle, and then file up bright all over; no use aiming for a forged finish as the yellow brazing will stand out like a sore thumb. If after this you find the two sets of bushes not in line, or the holes not parallel, bend or twist as needed to correct.

Now for the second method, which takes a bit more time but is more fun (at least, I think so). Start with a piece of mild steel $\frac{5}{32}$ in. $\times \frac{1}{4}$ in. (You may have to file down a bit of $\frac{1}{4}$ in. sq.) Make sure it IS mild steel—the first one I made was carbon (silver) steel and I finished with a U-shaped cutting-tool! Carefully slit down the centre, the slit about 2 in. long, and I recommend you put a $\frac{3}{64}$ in. hole through at the point where the slit is to end. The raw piece, by the way, should be 4 in. long, to give sufficient to hold it by. Get it red hot and open out the two legs, just bend them out. Put a couple of bits of steel angle in the vice to act as smooth jaws, and adjust the opening so that the unsplit leg will just go in. Heat at the junction, get it really hot, drop in the vice, clamp up and hammer the inside of the junction till the two legs stick out flat. This has to be done quickly, and will need several heats, as the vice will cool it rapidly. But *don't* be tempted to go on hitting the work after it is cold.

Now rough file the inside ragged faces where the sawcut was, and then heat these legs and forge them down to $\frac{5}{64}$ in. thick, with an occasional tap on the sides to keep the width to $\frac{5}{32}$ in. Use a piece of just under $\frac{1}{2}$ in. bar in the vice, heat up, and bend the 'U', just as described above. You now have a thick shank and a thin 'U' so the next step is to thin the shank. This can be reduced by filing or forging or both; I took off most by filing but finished with the hot forging, to get the appearance. Fig. 30 shows the various steps; you will note I have shown an enlargement at the junction, this is

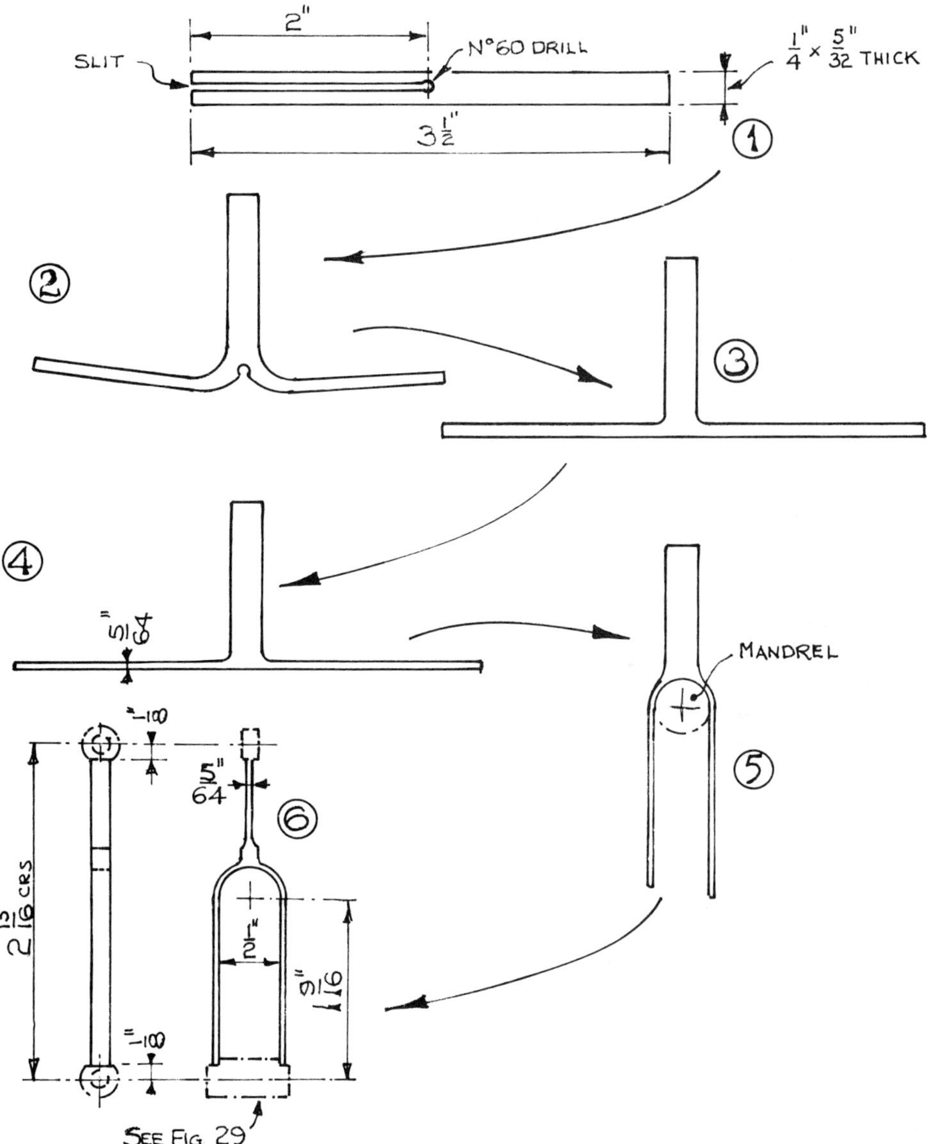

Fig 30- Steps in making the connecting rod, method 3.

 1. Drill small hole and cut slot.
 2. Open out arms—done hot.
 3. Hot forge junction of arms to flat.
 4. Reduce thickness of arms—file or forge.
 5. Bend round a mandrel—done hot.
 6. Bring to size and silver-solder the bosses in place.

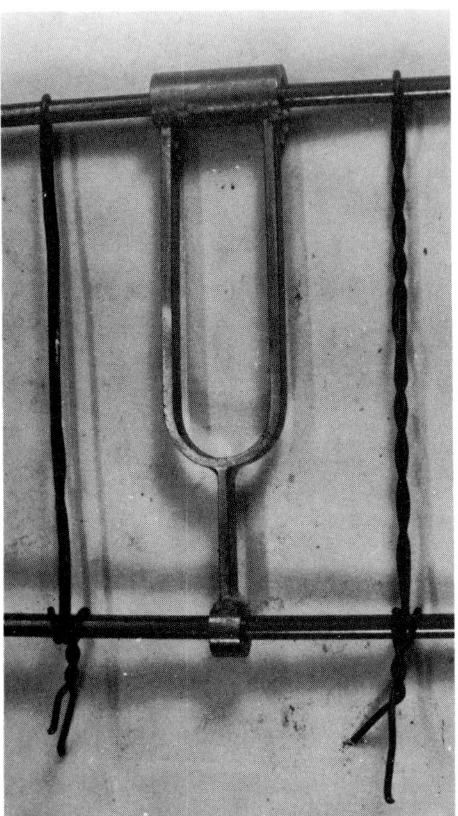

Fig 31 Connecting rod parts assembled and wired ready for silver-soldering.

optional, though likely to 'arrive by accident' as it did on one of mine. I'm not sure it isn't an improvement—strengthens the job and looks right. You now have to make two-bosses-in-one for the lower end as before, and a single boss for the top end. File flats on these and then offer up the at present over-long forging to these and cut off to make the overall dimension to $2\frac{13}{16}$ in. Note that the $1\frac{9}{16}$ in. must be held to $+/-\frac{1}{64}$ in. as otherwise the rod won't clear, and the overall centre-distance likewise as this controls the piston clearances in the cylinder. Fit up the bushes after filing the minimum of scale off for the joints as shown in Fig. 30, flux and braze up. As before check that all is square and bend or twist as neces-

sary. In this second method you can leave the job forged finish and just touch any bright brazing with matt black paint to disguise it.

Whichever method used of these two, you now saw out the un-needed centre part of the lower bushes and file the inner faces to dimension. Make and fit the three bushes as described earlier, when dealing with the bridle. Fig. 32 shows the finished rod.

Crosshead (Fig. 33)

This is made from $\frac{1}{4}$ in. $\times \frac{5}{16}$ in. mild steel. Chuck the bar in the 4-jaw, offset by $\frac{1}{16}$ in. so that the hole comes to $\frac{1}{8}$ in. from three of the faces, drill right through 7BA tapping size, face and bevel the end, and part off. If you use the 'marked jaw' procedure mentioned earlier you can reverse in the chuck to form the other bevel. Next, leave two opposite jaws as they are and open the other to grip the work so that you can drill and ream the $\frac{1}{8}$ in. hole for the piston rod. Make sure the work is square. This part of the job may be better done in the drilling machine if you have a vice which holds really square. Finally turn through 90° and form the boss for the other 7BA hole and drill and tap this. Clean out the $\frac{1}{8}$ in. bore and countersink the two opposing 7BA holes a trifle.

The two connecting-rod pins (B) are made from 7BA hex steel; if you haven't any, use the nearest you can get, or file up the hexagon from round stock afterwards, Turn down to $\frac{1}{8}$ in. dia., a good fit to the connecting rod bushes, then turn down again for 7BA $\frac{1}{8}$ in. long. Use the tailstock die-holder to thread this part. You need the thread to go as near to the shoulder as possible, and this is one of those cases where I use the reprehensible method of reversing the die in the holder (after cutting as much thread as possible) and cleaning up with that. Part off, reverse in the chuck and form the bevel on the hexagon. Make two of these. The 7BA set-screw (C) is simply a 7BA hexagon head screw cut short and with the head filed to square. Offer up the

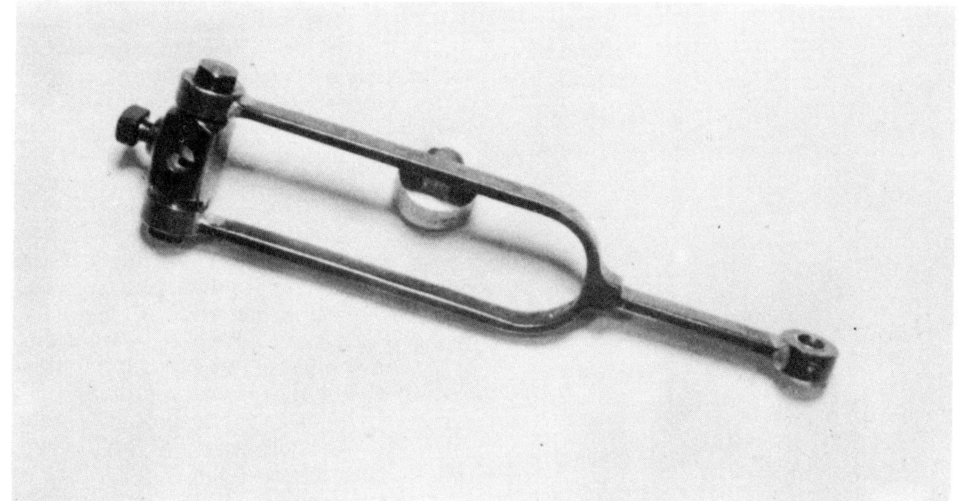

Fig 32-The finished connecting rod and crosshead.

crosshead and the pins to the connecting rod and ensure that nothing binds. Then put a piece of $\frac{1}{8}$in. ground rod through the rod hole of the crosshead and check that this and the connecting rod are reasonably parallel. If not, it's a case of a bit more bending and twisting. If, however, it is clear that the rod-hole in the crosshead is definitely out of square with the two 7BA holes in the sides it's best to make another—there is no way you can correct this error by bending the rod and there is risk of it binding as the crank revolves.

Base (Fig. 1 p. 10)

If you have already made a temporary base you can transfer the position of the column holes from this; otherwise use the dimensions shown on the General Arrangement drawing. (Correcting these as required if your arches are not quite to drawing.) Make sure that you get the drilling done before making the decorated rim if you are going to use the built-up base, using brass in that case, and steel if you are using the wooden base. (Whatever you do DON'T use aluminium; without doubt 5 years from now you will

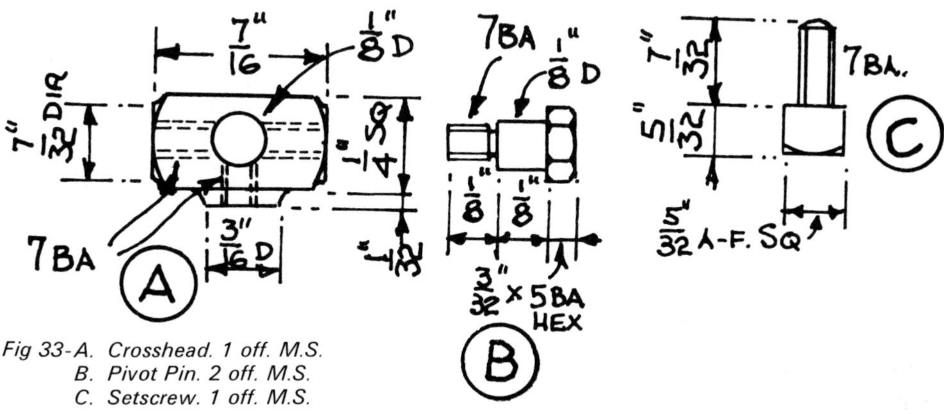

Fig 33-A. Crosshead. 1 off. M.S.
B. Pivot Pin. 2 off. M.S.
C. Setscrew. 1 off. M.S.

find corrosion under the paint around every hole.)

Mark out the two main centrelines and square up the sheet each and every way; then you can use your square on all sides without fear of error. Mark out for the four columns, drill, and erect these. Fit the crank and adjust positions till it runs free. (You should expect it to be a little stiff, but not to bind. It will free a little as it runs in; as a guide, it should take no more than 1 lbf force on the crankpin to cause easy rotation.) If not already done, make the elbow for the exhaust pipe—though these are obtainable ready made from Stuart Turner, etc. Assemble the cylinder set, making joints from stout brown paper or the proper 'Oakenstrong' jointing. No need to fit the valve at this stage, but pack the piston. The usual stuff is graphited asbestos but recently I have tried some PTFE yarn from Messrs Reeves. I can't speak for its wearing qualities, not yet, but it's as slippery, if not more so, than the graphite stuff and has the special merit that it doesn't exude a black 'goo' when running. A great advantage with exhibition models! Unbraid the yarn and pack it in thread by thread two turns at a time until the groove is full. Enter it to the cylinder—if it won't go in, roll the piston on a flat surface (lathe bed again!) until it enters. Apply a little oil and work back and forth a while. Pack the glands likewise, but don't overtighten.

Put the crosshead on the piston rod, no connecting rod as yet, and slip the bridle over this, right way up. Bring the cylinder to place and attach the bridle to the columns with temporary 8BA screws and nuts. Adjust the position till the rod slides up and down without binding or rude noises. (It's quite a musical instrument, this bridle, at times!) Clamp the cylinder set in this position. Drop the connecting rod over the bridle and assemble it to the crankpin. Fit the pivot screws and bring the crank to bottom dead centre. Push down the piston till it touches the cylinder bottom and then raise it a shade over $\frac{1}{16}$ in., (the design clearance is $\frac{5}{64}$ in.) and lock the setscrew. Rotate to top dead centre and make sure the end of the rod doesn't foul the 'U' of the connecting rod. Check that there is clearance at the top of the stroke, too, though it's hardly likely that you will be that far out!

All these checks having been made, mark through the bottom cover holes for the holding-down bolts, and also project down from the 'auspuff' (exhaust pipe to you!) and dismantle all. Drill for the four 6BA holes, using a size larger clearing drill than usual, to allow for some final adjustment, and a $\frac{3}{16}$ in. or $\frac{7}{32}$ in. hole for the exhaust. Disburr all holes—and I recommend a fair bit of countersink on the exhaust hole, to avoid the tendency for paint to peel at sharp corners.

Now assemble the cylinder set fully (Fig. 34). Check that the valve is free to lift off its seat and if not either reduce the width of the valve-nut or deepen the slot in the valve. Attach the valve-rod head with Loctite—it needn't come off again. Assemble the eccentric strap and rod, making sure that the 'pip' doesn't bind in the groove. Pack the valve gland and check for binding; you may have to open up the hole a trifle. Put the eccentric on the crank, boss away from the arch, and fit the strap and rod. Place the cylinder aligned with the holes (drop the odd bolt in) and position the eccentric so that the end of its rod just aligns with the slot in the valve rod. Measure for the washer between it and the arch. Make this washer, an easy fit on the shaft and about $\frac{7}{16}$ in. O.D. Assemble all (Fig. 35), with the steam-chest cover off. Don't connect up the valve. Check for free rotation, of course; you have to do this every time you dismantle, inevitable with 'so much lack of location'! Adjust the valve rod so that the valve nut is in the middle of the thread and connect up the eccentric rod to the fork. Very slowly rotate and make sure the valve clears each end of the chest. (It's about $\frac{1}{64}$ in. each end on mine.) If not, adjust till it does, and if it fouls *both* ends take off the steam chest and file both ends till it clears. (Don't file the valve. Stupid? Maybe, but it has been done, scores of times. Not everyone knows as much about steam engines as you do!) Now adjust the position of the valve so that the ports are open an equal amount at the end of the valve travel.

To time the valve, fit a brass grub screw in one eccentric hole and a longer screw in the other. Rotate the engine in the desired 'forward' direction; a matter of opinion, but when I was designing 'real' engines I always made them go round clockwise looking at the cylinder end (anticlock looking at the flywheel)

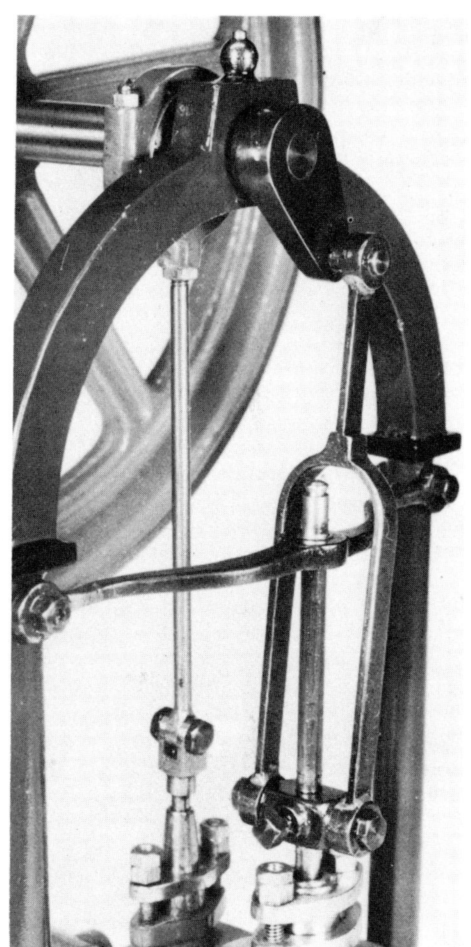

Fig 35-Close-up showing crank, bridle, and motion-work. Note the dummy lubricator.

probably for no other reason than that this was the way I wound up my then motor-car. (Which, for the record, cost me £3, with two gallons of petrol thrown in for goodwill!) So, if you are looking at the steam chest, as you must, to time it, rotate anticlockwise. With the crank on top dead centre, use the long screw to rotate the eccentric till the valve *just*, only just, opens the steam ports at the top end. Tighten up the brass screw, and rotate the engine (still the same way) to bottom dead centre. If the ports are visibly *open*, undo the eccentric rod and rotate the valve rod to bring the valve *down* by about half this opening. If the ports are fully closed, do the reverse. Then readjust the eccentric to get the ports *only just* opening at each end of the stroke. Now, you may not be able to 'equalise the lead' as we call this process, exactly. If not, set the greater lead at the bottom dead centre position. Put a drop of oil on the valve, and replace the chest cover. Don't forget the joint paper. Tighten the long screw, but don't fit the proper grub-screws yet.

Now for the moment of truth. Screw the end of a short length of $\frac{5}{32}$ in. O.D. pipe for the exhaust, and another to fit the inlet. Apply compressed air—or steam—after oiling round, of course, and see what happens (she will need about 10 lb. sq. in. at this stage). If the piston rod just jerks up to dead centre and stops there, fit a grub-screw to the flywheel and tighten it—one spoke in line with the crank. With any luck she will run and, as things ease up a little, pick up speed. Apply finger pressure to each arch—if she speeds up this is an indication of a little misalignment. Apply *gentle* finger pressure to the bridle—not much, as it's very delicate; the same applies. A little adjustment may be needed next time you assemble, but so long as she is running happily leave it for now. If the valve is correctly set, the beat will be reasonably even but if it is very irregular—CHUFF-chff-CHUFF-chff—then shut down and experiment with *very small* changes in the eccentric setting. Once satisfied, fit the proper grubscrews, one at a time, and tighten them up. Let her run as long as you like, not forgetting to send a shout for the family to come and see—nor the oil-can, either! After a while you should be able to reduce pressure and finally get her ticking over at about 100 rpm on perhaps 3 lb. sq. in. with the control valve almost closed. An odd thing here, you may find the bridle perfectly quiet at higher speeds, but giving a little groan or squeak every rev at slow speed. This disappears after some running, but can be cured by *very delicate* adjustment of the fixing screws.

Plinth (Fig. 1 p. 10)

I can't offer a great deal of help over the wooden base, as I am not all that good at this sort of work—though since my wife gave me one of these patent mitre-cutting saw-guides I do a lot better. However, a few lessons learned from making one of mine. First, don't use oak. The steel—in the screws and base—will be attacked by the acid in the wood; the former will rust and the latter stain. Mahogany was used a great deal by the Victorians for furniture, and they liked it just as well for engine plinths. Rosewood is another useful wood, hard and takes a good polish. Ebony is 'posh' (I made one of this, with little ivory feet) if you can get it. Brazilian Rosewood, sold to the school's woodwork classes for making recorders and such is superb. As to the joints, I used mitred ones, glued together with Cascemite, two screws in the base-plate each long side and one in the middle of each short one. One point; my drawing shows the plate right at the edge of the moulding. I recommend about $\frac{1}{8}$ in. inside this edge—allows a little tolerance on the woodwork. As to the moulding, I have a vintage moulding plane and used that, but enquiry at the local picture-framers revealed that it should be possible to get a moulding that will fit from such a source. Once made, drill the hole in the side for the exhaust to pass through.

For the built-up brass base, first reduce the pieces of $\frac{1}{2}$ in. $\times \frac{1}{2}$ in. brass angle to $\frac{5}{16}$ in. $\times \frac{7}{16}$ in. as shown. No reason why you shouldn't leave the $\frac{7}{16}$ in. as the original $\frac{1}{2}$ in, which will make it easier if a little tall. Cut this with mitred (45° joints), and to be on the safe side a shade short so that when offered up the baseplate projects just a shade all round. To get the joints dead square I used a carefully squared block of wood about 3 in. $\times$ 3 in. $\times \frac{1}{2}$ in. and cut off one corner just enough to be able to get the soldering iron in, and clamped to that. Don't braze it—distortion is almost certain. Having made the frame of angle, clamp this to the underside (don't get it wrong!) of the baseplate and tack-solder it in a couple of places each side.

For the next step I used a 100-watt electric solder-bit. The trick here is to warm the plate—not too much, but enough to give some expansion of the edges so that as you solder you don't set up strains. Mine fizzed when I spat on it, but wouldn't melt my resin-cored radio-type solder. Run a bead of solder round the inner junction of the angle and plate, working from centre towards the corners. This reduced the risk of unsoldering the corner joint before the placed solder had set. Go all round. Oh—naturally, it's best to tin both angle and plate first, and it helps to tin the outside of the angle too, to help when fitting the beading. This inner joint doesn't show, but nevertheless make as neat a job as you can.

I didn't use beading on mine, but slit a piece of tube—I have a little slitting saw on my small Boley lathe that does this a treat. (But the main reason was that I hadn't any $\frac{1}{2}$ in. beading.) Cut the beading down the middle, and file smooth. Note, it's *not* half-round, but a sector of a circle in section; most material suppliers stock it. Tin this, and then tack it to the angle. It's a bit fiddly, as you must mitre each end and offer up repeatedly till all four pieces fit. It It isn't really necessary to get a full-strength solder-joint on this part, as it is only cosmetic and any part of the joint not soldered can be paint-filled. However, by using resin-cored radio solder (which has a lower melting-point than tinman's) I was able to get a good joint all round on the one engine I made with this base. I did have to fill the odd gap at the mitre joints of the beading, though.

Oh-oh! My memory! Obviously, it's best to drill the hole for the exhaust pipe in the angle before soldering up, then all you have to do is to file the beading to clear the pipe.

Once the soldering is finished, file round the base to make it flush with the angle, and clean off all flux. I usually scrub such work in the sink with Stergene and very hot water but use your usual method. Finally, give a good rub all over with medium coarse (or medium fine— same thing) emery, to give a key for the paint. Fig. 37 (later) shows the wooden plinth; that in the heading photo is built up as described.

Painting

I don't as a rule deal with this, as there are books on the subject, and no two people seem to agree! However, my own practice is to spray all over with first red and then grey aerosol primers—the motor-car stuff. I then fill with primer-surfacer on any cast or not-smooth surfaces, applied by brush. This is left over-night in a warm place and then rubbed down with (on a small job like this) flexible-back No. 360 wet-and-dry silicon paper till the first coat of primer shows more or less all over, perhaps with patches of the grey or even metal and the depressions filled with the paler surfacer. Spray again with grey and then a second coat half-an-hour later. From there on I use either an air-brush or an aerosol, depending what colours I have available for the former. The aerosols are expensive, but it now seems impossible to get cellu-lose in small quantities, and I find (for my work) oil-based enamels too slow, as they seem to need too many coats to get a good finish when let down with thinners for the airbrush.

For this engine I would suggest the baseplate grey or black; the columns green, with the square ends picked out in black; and the cylinder and flywheel red, though the last could be the same colour as the columns. This fine work is ideal for the air-brush, but you do need a number of coats. For the cylinder I was fortunate in having some red 'Brushing Belco' so I first brush-painted this two coats and then rubbed down with No. 420 paper, finishing with two coats of the same paint let down with 50% thinners in the air-brush. All parts are then left on the top of my night-storage heater for two days, and finally polished with Brasso— the proper 'rubbing down compound' is too fierce for little jobs like this. The result is a gloss finish which is more or less 'scale paint' compared with a full-size engine when new. (Our full-size engines often had as many as six rubbed down coats after the initial filling, followed by the beautiful 'final enamel'—hardly practicable at current wage-rates!) After painting, take great care in reassembling the bridle; it is very easy to chip the finish when dealing with this part. I used washers under the nuts in most places, which again reduces the risk of paint chipping—not needed, of course, where a spot-faced surface is used.

Extras. The Feed-pump (Fig. 36)

This isn't shown on the main drawing but is detailed here—it was, in fact, an afterthought! It is a dummy as shown but can easily be made to work if you fit a valve-box under the base. However, the amount of steam used is very small and you would need an open bypass most of the time. The body (A) is turned from a piece of $\frac{3}{4}$in. diameter brass bar, but could be made from $\frac{1}{2}$in. stuff with flanges brazed on at the ends. The eccentric sheave (F) is the Stuart No. 7 engine casting, in iron, and the strap can be made either from a piece of $\frac{3}{16}$in. brass plate or, as I did, by cutting off the unwanted arm from the eccentric strap casting for the Stuart 'Launch' engine. (Stuart Turners will supply these at very reasonable cost.) To deal with the pump body first, chuck the material, say $2\frac{1}{4}$in. long, face the end, drill $\frac{1}{8}$in. about 1 in. deep, and with a knife tool form the $\frac{1}{4}$in. dia. boss—this is later screwed for the fixing, but don't thread it now. Put a small centre in the hole with a slocumbe bit, bring up the tailstock to support the end, and feed in the parting tool to form the width of the bottom flange; go in about $\frac{3}{16}$in. Reverse in the chuck, face the other end to correct length, and drill and ream $\frac{3}{16}$in. to the depth shown. Again, show this hole to a slocumbe drill and put in a small 60° bearing. Withdraw from the chuck and grip by the $\frac{1}{4}$in. dia. peg, supporting the other end from the tailstock.

Machine the body to shape. Whilst in the lathe, mark out the centres of the bolt-holes—*don't* drill them till you have made

The gland (B) is made from similar material; chuck in the 3-jaw, turn down the parallel part, drill $\frac{3}{16}$in, and form the 120° cavity with the point of a larger drill. Part off to make a flange $\frac{3}{32}$in. thick—leave a trifle on for later facing, which now do after reversing in the chuck. Mark out for the two holes. Remove from the

Fig 36-'Georgina' fitted with a feed pump. This engine has a Mahogany plinth. Compare with the photo shown on page 8 and note that in this case the cylinder has been used 'the other way up'. A slightly longer valve-rod is needed.

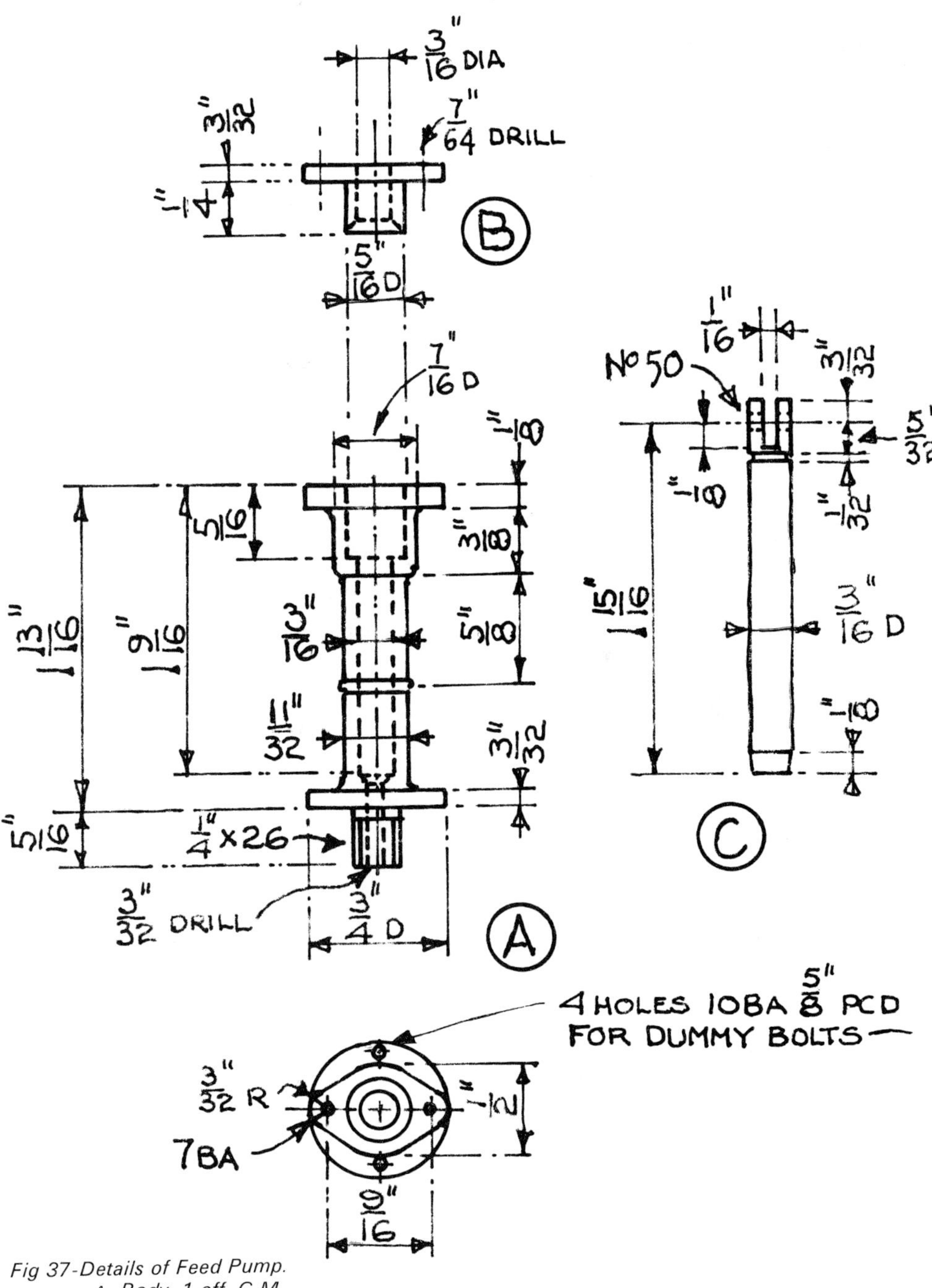

Fig 37-Details of Feed Pump.
A. Body. 1 off. G.M.
B. Gland. 1 off. Brass or G.M.
C. Ram. 1 off. Bronze.

See also pages 50 and 51.

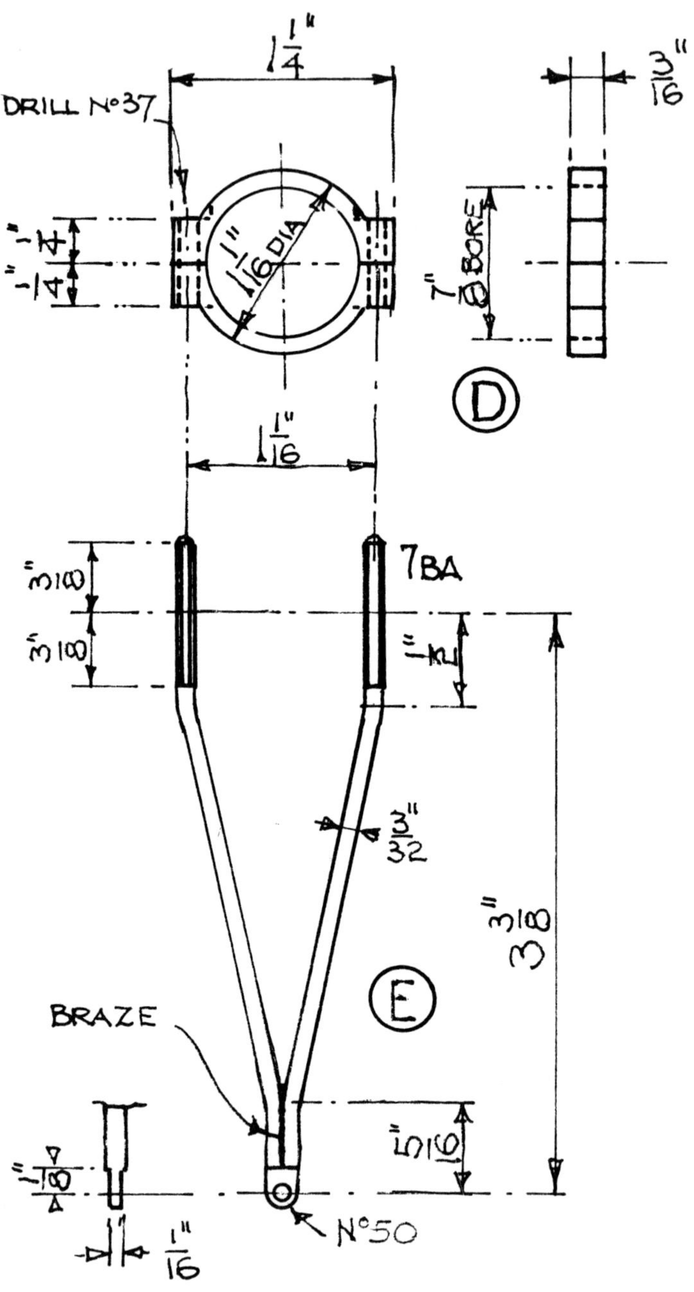

Fig 37- *D. Eccentric Strap. 2 halves. Brass or G.M. (See Text).
E. Eccentric Rod. 1 off. M.S.*

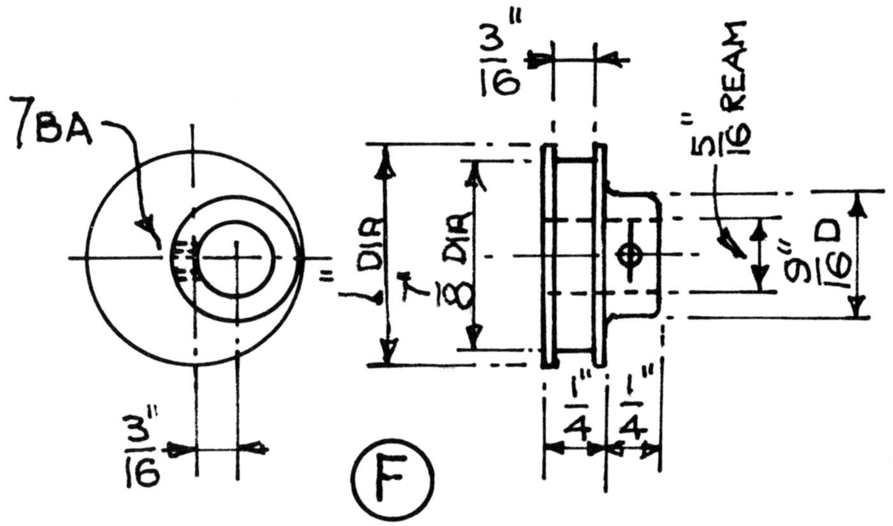

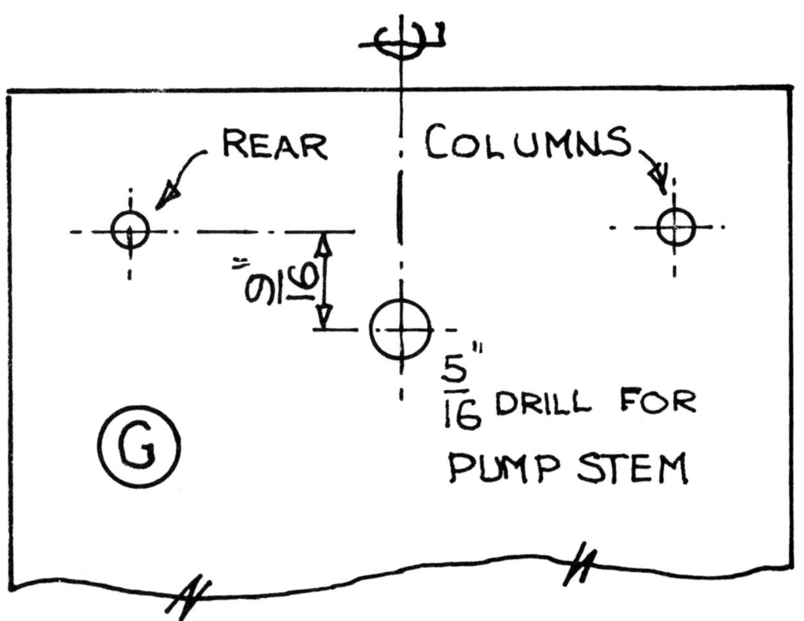

Fig 37- F. Eccentric Sheave. 1 off. C.I. Stuart No. 7, or M.S.
G. Location on baseplate.

lathe, centre-pop, scribe little circles to act as a guide in forming the oval and file to the line. Drill both holes, assemble into the body, pop through, and drill and tap the body. File the outline of the oval on the body, noting that this necessarilly will be slightly fatter in the belly as it were.

There are four holes 10BA shown in the bottom flange; these are dummies also, the pump being held down by the $\frac{1}{4}$ in. dia. part. Mark out for these, two holes being in line with those in the top, the others at right-angles. Tap 10BA. Finally, screw the $\frac{1}{4}$ in. dia. peg with any thread for which you have a nut; mine is $\frac{1}{4}$ in. BSF. This part can now be painted, and left to set whilst working on the rest. After which, fit four 10BA hex. head bolts, saw off the surplus below the flange, and file flush.

The eccentric (F) is made in the same way as that previously described, except that this one is of the flanged variety. This makes no difference, but you must work to a close micrometer dimension when machining the largest (1 in.) dia-meter, before forming the groove. Measure this exactly; then bring up your parting tool (run slowly) till it just touches the surface. Stop the lathe and make a little calculation to find out how deep to feed in the tool to get the rubbing surface the correct diameter. Plunge in to within 1 thou. of this figure, making cuts side-by-side till you have the width of $\frac{3}{16}$ in. Then feed in that extra thou. and traverse the parting tool from side to side to machine a smooth bottom surface.

Pump Eccentric Strap (Fig. 37)

I have included a photo of my engine at Fig. 37 which may help you to follow the drawing better. There is no eccentric rod as such, merely an extension of the eccentric strap bolts, which are larger than would be needed ordinarily. If you are using the suggested casting, saw off the cast-on gunmetal eccentric rod and trim up the casting. Saw the remainder across the middle—you will notice the hole is oval—and file the faces to bring the ears which carry the bolts almost to

size. Make sure you file flat. Solder the two halves together and then rough file the two faces to clear off any lack of flat-ness. I recommend you drill for the bolt-holes now, to give added security when machining. As there is plenty of metal you can do this in the drilling machine, but note you will have to spot-face for the nuts and bolt-heads too. Fit bolts and then grip in the 4-jaw and face one side. Reverse and tap down onto packing, with enough projecting from the chuck to enable you to face and at the same time keep the bolt-holes in the middle of the two faces. As before, don't grip too hard, as you are going to bore out most of the metal! You will have to bore to a 'caliper transfer' dimension if you haven't a ver-nier gauge narrow enough to measure the diameter of the eccentric between the flanges; set your outside calipers to the eccentric, and then, when boring, test your inside calipers against the outside ones. This is a normal procedure, used for at least 200 years after the invention of the steam engine, until micrometers got so cheap they sell them in D.I.Y. shops! Having bored, just touch the face with a knife-tool to give a circular outline to file the outer part of the straps to. With a scraper, lightly bevel the corners of the bore and relieve the said bore at the joints, as recommended when making the main eccentric. Remove the bolts, mark the two halves for later mating, and unsolder. Clean up the joint face.

For the 'rod' (E) you need two pieces of $\frac{3}{32}$ in. steel exactly $3\frac{7}{8}$ in. long. Use the tailstock die-holder to thread one end of each by a length equal to the height of the ears of the eccentric plus the thick-ness of two 7BA nuts. I can't give an exact dimension, as the latter vary. Mine came to $\frac{3}{4}$ in. At 1 in. from the end—the screwed end—put an equal bend in each rod such that the rods when fitted into the straps will cross their legs at about $\frac{3}{8}$ in. from the plain end. Put another bend in here to bring them side by side. Put nuts on the screwed end to hold them in the strap and adjust the bends till the lower ends lie truly side-by-side and the angles are symmetrical. Apply flux to the lower end and braze together, plenty of alloy, to fill the gap well.

File the sides of the joint to make it $\frac{1}{16}$ in. thick to drawing and then drill the hole as shown. The distance from this hole to the centreline of the eccentric strap is not critical, as there is plenty of clearance in the pump, but you should be within $\frac{1}{16}$ in. of this dimension if you have followed the drawing and these instructions. Round off the end of the rods.

The pump plunger (C) needs little work; true the ends in the lathe and machine the groove round it. Cross drill, using a jig if you have one for this size, otherwise just take care. Push a piece of $\frac{1}{16}$ in. rod through the hole to act as a guide whilst you saw down the slot at right angles to the hole. Widen this with a thin warding file till the eccentric rod is a slack fit. The knuckle pin is no more than a hexagon head 10BA bolt preferably with a plain section where it mates with the rod. Pack the gland lightly, the object of this being merely to provide a supply of lubricant as the pump does no work. It is quite in order to use brass studs and nuts on a water-pump by the way.

Set the eccentric sheave on the crankshaft, throw in line with the crankpin, attach the strap and rod and the pump-ram. Let this hang down vertically, and the end of the ram will point to where the hole in the base must be; arrange this so that the eccentric boss is a few thou. clear of the arch. You will, of course, have done all this before making the base in the ordinary way, but if you haven't, then you will have to dismantle and drill the painted plate. I had to do this, for as I told you, the pump was an afterthought! To protect the paint from drill-chips, stick on a piece of wide masking tape with a $\frac{3}{8}$ in. hole in it in way of the hole in the plate; this will save any serious scratches. Erect the pump body to the base, with a thin paper washer under the flange and tighten up the nut so that the long axis of the gland runs athwartships to the crankshaft. Adjust the position of the eccentric to suit—this should be just a whisper of clearance between it and the arch bearing. Attach the pump ram (after entering it to the bore) and check that it doesn't foul on bottom centre—though you will be a way out somewhere if it does! and tighten the eccentric grub-screw. You will need a whiff more steam to drive her now, but this will ease off as the parts bed in. As this eccentric is not used for timing it can be fitted with a trace more clearance than you would use for a valve but don't make it sloppy if you do have to ease it. Any real stiffness is more likely to be the gland not tightened up equally on both bolts. (This applies to the cylinder and valve glands, too.)

So, there it is. I think this addition to the range of engines from 'The Age of Elegance' has been worth while; certainly people who normally just take a passing glance at model engines have been rather more enthusiastic about this one. I shall look forward to seeing a few— preferably running at not more than 75 rpm!—at exhibitions in the next year or so. She needs a case? Of course! Messrs VISIJAR LTD of Gainsborough, Lincs, can supply one, say $7\frac{1}{2}$ in. $\times$ 8 in. $\times$ $10\frac{1}{4}$ in. tall, which gives adequate clearance all round.